Migration and development: a global perspective

Ronald Skeldon

 LONGMAN

Addison Wesley Longman Limited,
Edinburgh Gate, Harlow,
Essex CM20 2JE, England
and associated companies throughout the world

First published 1997

British Library Cataloguing in Publication Data
A catalogue entry for this title is available from the British Library

ISBN 0-582-23960-5

Library of Congress Cataloging-in-Publication Data
A catalog entry for this title is available from the Library of Congress

Set by 3 in 10/11 Plantin
Produced through Longman Malaysia, PP

Contents

Acknowledgements

We are grateful to the following for permission to reproduce copyright material:

Figure 3.1 from J C Chesnais *The demographic transition: stages, patterns and economic implications* (1992) The Clarendon Press by kind permission of Oxford University Press; figure 3.2 from J Salt & H Clout (eds) *Migration in post-war Europe: geographical essays* (1976) Oxford University Press by kind permission of Oxford University Press; figure 1.1 from W Zelinsky 'The hypothesis of the mobility transition' in *Geographical Review* (1971) Volume **61** (2) by kind permission of The American Geographical Society; Table 1 from P Bairoch *Cities and economic development: from the dawn of history to the present* (1988) University of Chicago Press by kind permission of The University of Chicago Press.

Whilst every effort has been made to trace the owners of copyright material, in a few cases this has proved impossible and so we would like to offer our apologies to copyright holders whose rights we may have unwittingly infringed.

Figures

Tables

Annexe tables

Preface

With sustained reductions in fertility either realized or under way throughout the greater part of the world, and mortality, too, now well in decline, it is migration that has emerged as the major population concern as we move into the twenty-first century. The trend towards a global community in which information and capital can be moved across state boundaries virtually instantaneously has profound implications for human mobility. While the movement of labour is certainly not as fluid as that of capital or information, there has unquestionably been a significant rise in the volume, types and complexity of human migration over the past few decades. As levels of development for many have risen, and differences in living standards from one part of the world to another have widened, so too has migration increased. The economic and demographic viability of certain communities has waned, while the fortunes of others have been invigorated. Such a mixing of different peoples and diverse cultures is taking place that the integrity of entire nations is being questioned. Central to these issues is the role of the city, both as a focus for the concentration of populations and as a key node linking international, regional and local economic and mobile systems.

The increase in migration has been matched by the growth in writings about the topic, and analyses from every conceivable point of view continue to pour out. Economists, sociologists, anthropologists, political scientists, statisticians, legal specialists and historians, as well as demographers, have all made their own special contributions. It is virtually impossible to be aware of, let alone dominate, the vast literature available. Yet, it perhaps falls to a geographer, trained to believe that geography is indeed one of the great synthesizing disciplines, to attempt to create a coherent and integrated picture of this complexity. Thus, the main purpose of this book is to chart a path through the voluminous literature on the subject and to provide a framework into which may be placed the various, and sometimes contradictory, findings so that these may be more clearly viewed and understood. The book builds on my earlier work, *Population mobility in developing countries: a reinterpretation* (London, Belhaven, 1990), but extends the coverage to the more developed parts of the world to provide a global picture. It also incorporates international as well as internal migration. Both this book and the previous work rely very heavily on a wide range of secondary sources, although there is relatively little overlap in the references used for the two books. With several notable exceptions, this book focuses on the literature published since the late 1980s, and readers interested in an assessment of the work on migration in developing countries published before that time should consult my earlier book. As far as possible, acknowledgement of the sources of ideas is given in the text and I beg the authors' indulgence if I have used their data and interpretations in too injudicious a manner.

This book had its origins at a lunch hosted by Professor Denis Dwyer when, in response to the innocent question 'Who is writing the book on migration in your "geography and development" series?' he replied 'You are', hence demonstrating the old adage that there is no such thing as a free lunch. The book was written during the first half of 1996 when I was on leave from the University of Hong Kong and a visitor at the Population Division of the United Nations Economic and Social Commission for Asia and the Pacific (ESCAP) in Bangkok. My thanks go first to Dr Nibhon Debavalya, Chief of Division, who facilitated my attachment to the Division. He and his staff provided a quiet environment that was ideal for writing. The regulars at the lunch table in the ESCAP cafeteria generated the witty breaks so necessary to keep the writer sane during the long hours of writing. At the Department of Geography, University of Hong Kong, Mrs Meimei Cheung performed her usual stalwart service, although this time from a distance, in entering the first drafts of the early chapters on disk. Until I finally learned the art of directly composing onto the screen, she responded efficiently to all the many packages of pages arriving from Bangkok. Once again, I am deeply grateful to my editor wife Grania, who invested much time, effort and concern in removing grammatical imperfections and textual inconsistencies that will have made the reader's task much easier than it would otherwise have been. To her we all owe an immense debt of gratitude.

This book represents a personal transition away from the paper rice bowl of an academic position at the University of Hong Kong, where I spent fourteen happy years, and I would like to thank all my colleagues there for the friendship and intellectual stimulation given over the years. It is now on to those 'fresh woods, and pastures new' and a transition that will certainly involve much further migration and all the developments that will ensue.

Ronald Skeldon
Bangkok, March 1997

Myths and movements

The fact that a book is being written about migration and development implies that there is some kind of relationship between them. That this is the case will be apparent, but the relationship is exceedingly complex and few simple, or indeed causal, linkages can be established. The first stumbling block is the semantic difficulty of defining the two terms. We all intuitively know what 'development' and 'migration' mean but, when we come to identify and delimit their substance precisely, they prove elusive indeed. Both are dynamic terms that imply change: development suggests a growth, an evolution, an advancement; migration suggests a shift in place of residence from one area to another.

Since the end of the Second World War, development has been a shibboleth of politician, planner and social scientist. Development gives the impression that humans are in some way in control of their destiny and can improve their condition. There is a clear idea of progress, that things are getting better. Unquestionably, conditions have improved for vast and increasing numbers of the world's population over the last fifty years in terms of their capacity to feed themselves, their length of life, their standard of health, their level of education, and so on. Yet, for large numbers of others, conditions have hardly changed and, in some areas, have deteriorated. In the late 1980s, it was estimated that over one billion people, or more than one fifth of the world's population, lived in poverty (World Bank 1990). In terms of absolute numbers, there are more poor people alive now than there have ever been, and this numerical fact is largely the product of past and continuing rapid rates of population growth. This is not to suggest that poverty is caused simply by demographic growth. It is more the result of political factors and the failure of human agency, even if the demographic trends have provided the foundation for the magnitude of the problem.

One of the characteristics of the poor is that they are found disproportionately in rural areas. Yet, the populations living in rural areas have declined relatively, if not absolutely, over the last forty years. At the same time as the population of the world has more than doubled, from 2516 million in 1950 to 5292 million in 1990, so has the world become a more urban place. In 1950, about one quarter of the world's population lived in cities. By 1990, this proportion had risen to 45 per cent, representing 2400 million people. Very generally, the most highly developed countries are also the most highly urbanized, although the converse is more exact: there is not one poor country in terms of either economic or social indicators that is highly urbanized. Hence, the transfer of population from rural to urban areas appears to be an integral part of any process of development, and one of the principal ways in which this transfer is made is by rural-to-urban migration.

However, any relationship between migration and development must consider more than a transfer of population from rural to urban areas during a transition to an urban society. Migration itself encompasses more than a simple unilinear movement between rural and urban sectors and needs to be conceptualized as a complex system of short-term, long-term, short-distance and long-distance movements that can better be subsumed under the term 'mobility'. Much of this will be neither permanent nor directed towards urban centres. Just how 'short-term', 'long-term' and 'short-distance' and 'long-distance' are defined is problematic and will, in practice, be controlled by the spatial and temporal units used by individual countries in their censuses and surveys. These vary from country to country, which makes direct inter-country comparisons extremely difficult. It is not just inter-country comparison, however, that is problematic. Longitudinal analyses within any single country are often complicated by the fact that the spatial and temporal migration-defining units vary from one census or one survey to the next. This book, which focuses on global issues of migration and development, is not the place for an extended discussion of the problem of the measurement of migration. Some of the main issues are raised in Chapter 2 and these have been discussed in detail elsewhere (Skeldon 1987a). Measurement remains one of the critical areas of migration research simply because the measurement of the volume and types of migrants can be no better than the definitions adopted in censuses and surveys. All people move during their lifetime and it is incumbent upon the analyst and policy-maker to decide which of these moves is significant.

Migration, or more exactly mobility, of some form is a universal experience and it is rare for anyone to spend his or her entire life within the boundaries of a single village or city ward and, when this does occur, it is more likely to be due to some physical or mental handicap than to choice. It is worth remembering that in virtually all societies a basic punishment is the withdrawal of the freedom of movement though imprisonment. Migration, technically, does not become 'permanent' until an individual reaches his or her death place, and even then, as we will see, it still need not necessarily be permanent. Although the word 'migration' will be used throughout this book, as it is after all a perfectly good word in the English language, readers must be aware that it includes a whole spectrum of human mobility. It is the variation in the forms and types of this mobility, and their relationship with development, that are the focus of this book.

Like migration, development too must be disaggregated into many parts. Some of these parts are related to technological change and here an association with population movement can be direct. For example, improvements in means of transport have allowed people to travel faster, further and more frequently. Yet, development implies much more than technological change; it involves institutional changes in society that have political expression. The forms of government influence the responses of those other powerful actors on the world stage that shape the international economy of the later twentieth century: multinational corporations and banks. The latter, in conjunction with the state, affect the global diffusion of industries, services and transnational flows of capital, which, in turn, affect flows of labour. It is the fundamental premise of this book that there can be no significant development without the emergence of new forms of spatial mobility. Population mobility is thus an integral part of the development process: it both causes and is caused by changes in the economic, social and political structure of an area.

It would be incorrect to see migration as either positive or negative for development: it is but part of that whole process of change that is implied in the term 'development'. Migration can be viewed as both positive and negative at the same time. For example, the remittances sent back by migrants can be a significant source of foreign exchange for a developing nation and critical to its development objectives, as in the case of the Philippines. Yet, that migration itself reflects a lack of development in the home country and may have negative consequences for families separated in time and space or for communities deprived of some of their most active and innovative young men and women. The immigration of large numbers of workers may cause the persistence of otherwise uneconomic forms of production at destination areas and thus could be seen as the antithesis of development. For example, the movement of large numbers of illegal migrants from China to be kept as virtually bonded labour in sweatshops of the Chinatowns of New York or San Francisco in a sense retards the development of these cities. Without the migration, the labour-intensive industries would be forced out because of labour shortages in these particular non-skilled sectors, and the cities could move faster to more knowledge-based production. However, the migrants themselves, despite their tenuous and exploited position, may be better off than if they had remained in their home villages. Even if they are exploited economically, they may have feelings of hope that they could not have had in their villages. In other cases, the migration of large numbers of workers from rural to urban areas could be seen as good for development as it leads to an equalization of wage levels, not only by slowing wage increases in the towns but also by increasing the flow of income to the rural areas through remittances. Thus, the implications that migration may have for development can vary according to the level of analysis, that is, individual, community, nation or state, at the microlevel or macrolevel.

There is an implicit and often explicit tendency to view migration and the migrant in a negative light. Our studies tend to be biased towards why people move, as if movement were somehow anomalous and a settled state were the norm. The counterfactual question 'Why do people not move?' might bring a very different perspective to our analyses and, ultimately, policy (Harris 1995). Philosophical musing aside, this negativity is reflected at the practical level of policy. Migration policies are rarely implemented to facilitate the free movement of people; they generally seek to control, regulate or limit population mobility. Internally, they are often designed to divert, slow or stop migration towards the largest cities in any country and, internationally, policies seek to restrict access to citizenship and residence by foreigners. There is an almost universal fear of the outsider that makes him or her an object of suspicion at best, and of abuse and exploitation at worst. The marked growth in the number of new states in the second half of the twentieth century, as an era of nationalism has replaced colonialism, has seen a multiplication in the number of barriers against migration to the extent that the term 'global apartheid' coined by Richmond (1994) seems singularly appropriate. Despite these restrictions, population migration has increased over the last four decades and the means by which people move have become more complex. Our era has indeed become an 'age of migration' as Castles and Miller (1993) aver, although perhaps not entirely in the manner envisaged by the authors of that estimable book. Certainly today, more people are moving in more ways than ever before.

The United Nations (1994) has estimated that, in the last half of the 1980s, somewhere between 750 million and one billion people migrated, and the numbers appear to be increasing every year; that is, at the global level, perhaps one person in every six migrated over that five-year period. It is magnitude that must be emphasized, however, as a mixing of cultures is not new in our history. Before the era of nationalism and the attempt to create homogeneous states, most cities and empires were a patchwork of different peoples. Our present era, in which nationalism is receding in the most developed parts of the world at least, is seeing the 'reassertion of the polyethnic norm' (McNeill 1986). The fear of migration is the fear of the new identities that will result, and that stems from a lack of understanding of our past. There are no ethnically 'pure' groups.

The question is not so much of numbers or of rates, however, but from where and to where the migrants are going, and who is moving. Most of those moving are in the poorer parts of the world simply because the greater part of the world's population is to be found there. It is a minority of the total number of migrants which moves to urban areas and an even smaller proportion of the total which moves from poorer to richer countries. Yet, these two flows, the internal migration to urban areas and the international migration to developed countries, cause the greatest concern. The concentration of population in megacities in many parts of the world and the transfer of people from poorer countries to wealthy developed countries with very different cultures are seen as potentially explosive issues politically, socially and economically. In part, this concern is again a fear of the unknown: urban areas larger than any yet seen are emerging, and a mixture of cultures of a magnitude never yet experienced in human history is occurring. The absolute numbers are indeed huge but for international migrants, who represent a small minority of the total number of movers, the rate of migration from poorer countries in 1990 did not appear to have increased since 1970 and the movement from industrialized countries had fallen (World Bank 1995: 53).

It is not just these international movements that must be of concern. There are important rural-to-rural flows, some to open up new areas of agricultural land in environments that may be marginal or fragile and subject to degradation, and there are increasing movements among poorer countries themselves. With the decline of global fertility by some 30 per cent over the last forty years, the 'population explosion' appears to be coming under control. There remain parts of the world where rapid population growth is still a problem but in many other areas persistent low fertility and slow demographic growth are being perceived as political threats to society and economy (Teitelbaum and Winter 1985). As we move into the twenty-first century, it is likely that the major issues to be faced in the population field will revolve around migration and population distribution rather than around high fertility and population growth.

No highly developed society is characterized by high fertility and while there is no simple correlation between fertility decline and any single development variable – per capita income, rate of increase of per capita income, education, urbanization, and so on – that decline has unquestionably been affected by development. Even where fertility is much lower than would be expected, given levels of economic development, as in Sri Lanka or Kerala State in southern India, regional societies have been modified through the introduction of health services and clinics and by non-traditional political and administrative structures. As forms of 'modernization' and development have been shown to reduce fertility, albeit in various ways, so too policy makers and analysts have considered that migration may be affected by modernization and development.

At the simplest possible level, it can be said that people generally move from poorer or otherwise disadvantaged areas to richer or more advantaged areas. Hence, will making the poorer areas richer reduce migration? Will development in its broader sense reduce migration? The historical experience of Europe shows that, at the macrolevel, migration increased during initial industrialization, suggesting that migration increases with development (Massey 1988). In addition, development in its broadest sense has increased the wealth differences between areas. Before the Industrial Revolution, wealth differences were limited, perhaps with a ratio between the richest and the poorest in the region of 1 to 1.3 or less (Bairoch 1986: 192). With industrialization, differences accelerated dramatically, with per capita wealth increasing perhaps 1600 to 1800 per cent amongst the leaders of economic development and not at all or even decreasing among the poorest areas. Considerable migration did characterize the pre-industrial world but so too did it typify the industrial and post-industrial worlds, though its form and function changed. The relationship between migration and development is complex and not direct, and migration is unlikely to be slowed by development.

Notwithstanding the historical experience, there are attempts to reduce migration through development planning and four general policy areas have been identified: trade, foreign direct investment, aid and the promotion of political freedom (Martin 1994: 241). Trade and foreign direct investment should lead to an increased number of jobs, while aid can be used to promote the kinds of policies that governments need to implement in order to achieve higher levels of development, as well as more direct ways of improving human welfare. Moves towards more openness in government, with greater participation of people in decision-making, should provide a political environment more liable to retain people than to force them to flee.

Of these four general areas, the one that is most amenable to direct intervention by the governments of developed countries is aid, but all the signs are that the volume of direct assistance from developed to developing countries is declining. It certainly declined as a percentage of gross domestic product (GDP) from 1960 to 1992, from 0.44 to 0.35 per cent of the GDP of developed countries (Stalker 1994: 162), even if absolute amounts increased. The relationship between aid and migration has begun to receive attention in the literature (Böhning and Schloeter-Paredes 1994) and, while conclusions are as yet tentative and mixed, it appears to be clear that aid can be but one supporting factor among the many domestic policies that countries can pursue in order to reduce the number of people wishing to leave. Whether it can actually reduce migration, even assuming that this is a desirable policy direction, is entirely another matter. Given that the amount of money sent back by migrants in remittances through official channels only, and hence a severe underestimate, is equivalent to over half the value of government aid (Russell 1992: 269), this seems a dubious proposition indeed.

This discussion introduces one of the nebulous terms that has become current in the migration and development literature, namely 'migration pressure' (Straubhaar 1993, Schaeffer 1993). What this is precisely and how it should be measured remain problematic. This term, like its parent, 'population pressure', is clearly trying to convey some idea of overpopulation: that in some parts of the world more than in others some kind of imbalance exists between population and resources which is likely to lead to migration. Migration pressure is the result of an excess supply of people willing to migrate relative to the demand for people in potential destinations (Straubhaar 1993). The analogy of a dam of migration controls holding people back and allowing the pressure to build up may be intuitively appealing, but the implication that the pressure will be released once the barriers are removed does not accord with the observed experience of migration. Migration leads to further migration rather than abating, and large flows can evolve from areas where there was no apparent pressure. The whole issue of pressure appears to imply ideas of balance, of equilibrium and disequilibrium, that may not be entirely appropriate for the analysis of the constantly changing conditions that are likely to affect population movements. Migration is not simply the result of a quantitative balance between numbers and resources, implying that it will be more pronounced where relative deprivation is greatest, which is perhaps the largely unintentional implication of the term 'migration pressure'.

At the outset, it is worth outlining five misconceptions that we could term the 'myths of migration', which still permeate much of the popular and indeed, and less excusably, the scholarly literature. Some are more prevalent than others but all need to be highlighted at this stage so that the reader is forewarned.

The first is the myth of the immobile peasant. The implication here is that migration in pre-industrial, traditional or 'pre-modern' societies was limited. Research on the historical experience of

Europe (see especially Moch 1992) and Japan or on present-day rural developing societies has shown that peasant societies are and always have been highly mobile. Mobility among certain groups was greater than among others and it was greater at certain times than at others, but migration was very much a characteristic of past and present rural societies.

An extension of this same myth is the view that non-European societies were somehow static or stagnant before being galvanized through their linkage to the European-dominated world system. Before, and for a long period after, European expansion had begun, brilliant civilizations in Africa and Asia were developing in their own way with complex systems of trade and human mobility. Complex patterns of migration were not simply a product of so-called modernizing influences diffusing from Europe. A global view of migration and development carried out a millennium ago might have concluded that we were on our way to a world dominated by China as the core of a global system. Yet, in the following thousand years, the centre of world power shifted gradually towards Europe and the Americas (see Fernandez-Armesto 1995). There was nothing inevitable in this transition and only in the Americas did indigenous civilizations succumb early and quickly to European advance. Elsewhere, that progress was uncertain and fitful until the nineteenth century, when the wide availability among European powers of the innovations of the Industrial Revolution gave them technological superiority (Adas 1989). It is surely unreasonable, then, to expect sclerosis in the world system now and to imagine that future shifts in power will not occur in the next millennium.

The second myth, although somewhat contradictory to the first, is that it is the poorest who move. The poor do move, but so do wealthier groups, and the latter are likely to move further and more often than poorer people. Thus, making the poor richer will not generally lead to lower migration. On the contrary, it will see the persistence of migration and usually, although not invariably, an increase in movement.

The third myth is that rural-to-urban movements represent the dominant type of migration in developing countries. This can be true, but only under certain conditions; generally, rural-to-urban flows are relatively small compared with flows within the rural sector, within the urban sector or even from urban to rural areas. Urbanization, or the shifting balance between the urban and rural sectors, unlike the growth of individual cities, has not been particularly fast in the developing world compared with the historical experience of Europe. As a corollary, we can add a fourth myth: that migration is the principal component of urban growth in developing countries. In some cases it is, but, more commonly, natural increase and the extension of urban boundaries are more significant components of urban growth.

The fifth myth is that migration, whether internal or international, is a simple move from an origin to a destination. It can be, but it is far more likely to consist of a complex sequence of moves that may involve several destinations and regular contact with the origin, which may eventually involve return migration. One of the common examples of this myth is in the idea of a European settler as compared with an Asian, usually Chinese, sojourner. The first implies a permanent family movement while the latter implies a temporary, mainly male, migration. As will be apparent in subsequent chapters, many Europeans were sojourners and Asians were settlers.

All these myths draw attention to the basic fact that migration is an exceedingly complex process and it is impossible to reduce its description to a few ready generalizations. In addition to avoiding the above myths, the arguments in this book will also seek to undermine two general conventions in the literature: first, that there is a clear division between internal and international migration and second, that there is a clear division between migration in the developed world and that in the developing world, a world which is still, and unfortunately, termed 'Third World' by many.

The division of migration studies into internal and international movements has probably had as much to do with the sources and types of data used to measure the two movements as to any substantive or logical differences between the two. There are some valid reasons to maintain a separation between internal and international movements as the crossing of, and the controls between, international boundaries in some areas are much more marked than in the case of internal administrative boundaries (see Cohen 1995b: 5, for example). Nevertheless, there are links between internal and international moves, and boundaries drawn between them are often artificial, bisecting integrated migration systems. Another important drawback is that two almost separate traditions have evolved and those researching in one part of the subject rarely have anything to do with those in the other. There have been relatively few attempts to integrate both internal and international movements within a single framework, although for an early example, see Pryor (1981) and, perhaps less successfully, Kleiner et al (1986). For a review of the historical relations between internal and international migrations in Europe, see Baines (1994).

In several parts of the world, particularly in West Africa, what are migrations within single ethnic or cultural groups became international migration simply because of the way in which European colonial powers constructed the administrative boundaries that later gave rise to independent states. In the more developed world it is more realistic to consider the migration between Ireland and Britain as internal, rather than international (Baines 1994: 37). In addition, the reduction of states into smaller units has reclassified what were once internal into international movements. The dissolution of the former Soviet Union into a series of independent states is perhaps the most obvious example. Hence, the distinction between internal and international flows can be one of administrative convenience rather than of any substantive difference in the nature of the flows. There are clear linkages between internal and international moves; this book will cover both and attempt to provide a framework in which to integrate both satisfactorily.

Associated with the need to consider internal and international movements together is the desirability of integrating developing and developed parts of the world within any overall discussion of migration. In part, this is a reflection of the current concerns with globalization which, in turn, have evolved out of world systems analyses. The basic tenets of this approach contend that, in order to understand the situation in any part of the world, it is first necessary to consider how that part is linked to the major global centres of power. These linkages reflect the increasing economic interdependence of the world through an evolving international division of labour in which each area capitalizes on its particular comparative advantage in the production of a good. Thus, international trade and international migration become linked.

In part, the need to consider developed and developing countries together reflects the increasing interchange of population between the two regions that has occurred over the last fifty years. In part, too, the need to integrate developed and developing countries together in any discussion of development emphasizes that the developed countries are also developing. There can be no intuitive end-point at which a country ceases to develop, particularly in the context of continuous change engendered through technological development. Of course, discussions of the implications that the ageing of human populations will have for the future welfare of the most advanced countries, and of the fact that future generations cannot expect to be as prosperous as the present one, suggest that there may come a point after which countries become 'undeveloping'. Such concerns lie beyond the scope of this book, which is concerned with economic and political change and human mobility irrespective of whether these imply progress *sensu strictu*.

However, the twofold division of the world into developed and developing is too simplistic. The demise of the Soviet Union and the end of the superpower rivalry with its international relations based around a tripolar world of liberal-democratic free-market economies, socialist centrally planned economies, and 'the rest' classified as the 'Third World', has engendered a search for a New World Order, which has yet to emerge from the post-Cold War firmament. The reorientation of the centrally planned economies towards western-type market forces has favoured dualistic terminology to describe the growing disparity between rich and poor; and 'North' and 'South' are seen by some as preferable to 'developed' and 'developing'. The increasing cooperation between the ex-First and ex-Second worlds has deepened the sense of isolation of the South as an area being left to its own devices within the context of the New World Order. The majority of countries making up the South were originally colonies of the North and the post-colonial experience provides a sense of common identity and feeling of exploitation. The North has its own powerful institutions, the North Atlantic Treaty Organization and the secretariat of the Organisation for Economic Co-operation and Development, for example, but the institutions in the South to promote South–South integration or to provide common fronts in negotiations with the North are weakly developed. The Group of 77, or the Non-Aligned Movement, is politically weak and participation in the United Nations and its specialized agencies has often proved frustrating as the interests of the North appear to dominate. To strengthen solidarity in the South, it has been recommended that a 'Secretariat of the South' should be established (South Centre 1993) that could adequately represent the South and coordinate self-reliance and people-centred development among the poor.

Such thinking is entirely meritorious, but it appears to deny much of the reality of both human history and the geography of the world. In Wallerstein's comments on the reports of the South Commission, he observes (1993: 118) that:

> An appeal of the liberals among the powerful to their compeers to make reforms in the interests of equity, justice, and heading off worse has never had any significant effect in the past several hundred years except in the wake of direct and violent rumblings by the oppressed, and it will have no more effect now.

The South too is a concept of nebulous geography. As an analytic category, it can lead to difficulties such as the following comment on Jean Gottmann's list of world cities: 'With the possible exception of Beijing, therefore, all Gottman's world cities are in the North ...' (Simon 1992: 186). Quite irrespective of its 40°N location, Beijing literally means 'northern capital'. The idea of the South is to convey powerlessness and poverty rather than geographical coordinates, but attempts have seldom been made to identify clearly which countries of the world make up the South or, more importantly, whether these countries have common aims or common interests. Just as some consider that the Third World ended some ten years ago (Harris 1986), the South as a meaningful construct appears moribund from the start as it papers over fundamental economic, political and cultural differences between the countries themselves.

An important component, however, of the new political landscape has been the remarkable development of a small number of economies, mainly in East and Southeast Asia, that have grown to levels approaching, and in the case of Singapore surpassing, those of many developed countries. China, in terms of average gross national product (GNP) per capita, still lags far behind but, given its sheer size and its links to a global network of dynamic overseas Chinese groups, some commentators see it as one of the major economic and political powers of the next century (Overholt 1993). The emergence of other, if less spectacular, growth nuclei in southern Africa and the Middle East, as well as the existence of stagnation or deterioration in other areas, renders any common categorization such as 'Third World', 'the South' or even 'developing economies' to cover all of Latin America, Africa and Asia virtually meaningless. Complex patterns of variable development now exist within these vast areas, as well as within the area covered by the former Soviet Union and, just as important, among what are generally accepted as the developed countries themselves. Chapter 2 discusses more fully a more elaborate regional framework for the analysis of migration and development. What needs to be stressed here is that, while a distinction between 'developed' and 'developing' countries is still a convenient generalization, and the use of these terms is not going to go away – I use the terms in this book, for example (although I do *not* use 'Third World') – intra-category variation is as important as any inter-category differences that can be identified, and the use of these terms is imprecise. A clear objective division of the world into two separate economic, political or cultural groupings is not a particularly useful analytic approach.

This discussion raises another misconception that can be added to the previous myths of migration, which is that the migration experiences of the countries that are commonly classified as developed have been totally different from those of countries that are seen as developing. To dismiss this view as a myth does not imply that the experiences are or have been the same across all countries. Rather, and following the previous argument, there is a whole range of migration experiences, but there are important similarities as well as differences cutting across countries at very different levels of development.

In a previous examination of migration (Skeldon 1990), I focused almost exclusively on internal migration and on movements within the developing parts of the world. The emphasis, as one reviewer observed (Findlay 1991), was on the search for order and, in that search, I may have overemphasized universal models at the expense of variation. This book will persist with a search for order. To accept that each variation is unique and complete in itself will lead us down the sterile path of relativism and return us to an exceptionalism that was all too common in geography in the past. The concern in this book is much more with accounting for difference – the variation-finding comparisons that 'help us make sense of social structures and processes that never recur in the same form, yet express common principles of causality' (Tilly 1984: 146). Tilly's other methods of comparison, the universal comparison based on general models, the individual comparison of specific cases and the encompassing comparison locating a process or phenomenon at several points in the same system, were examined in the previous work referred to above (Skeldon 1990). The present book, while not ignoring these approaches, will attempt to account for the variations in migration from one part of the world to another, encompassing both internal and international movements and covering both more developed and less developed parts of the world.

My task is thus much more ambitious than in the previous work in terms of coverage, particularly so given the volume of literature on the subject of migration. In the late 1980s, I observed the outpouring of books and articles on the topic but, by the mid-1990s, this has become a torrent. We may be in an age of migration; certainly we are in an age of writing and talking about migration. One commentator has observed that, in the first half of 1993, some eighty to ninety intergovernmental meetings were held on migration issues in Europe alone (Spencer cited in Black 1995: 267). In the last few years, several excellent appraisals of aspects of migration at the global level have appeared: Castles and Miller (1993), Parnwell (1993), Potts (1990), Richmond (1994), Stalker (1994), Stark (1991), Weiner (1995), Zolberg, Suhrke and Aguayo (1989), and these have been supplemented by many journal articles and edited collections culminating most recently in the massive *Cambridge survey of world migration* (Cohen 1995b). As this book was being written, there appeared Sowell's *Migrations and cultures* (1996), Weiner's review (1996), two collections of essays on global migration (van den Broeck 1996; Wang 1997) and a multi–volume compilation of previously published recent articles on migration (Cohen 1996b).

One might legitimately ask what else could possibly be written on the topic. Surely there must come a time when one's contribution to scholarship is not to publish, the logical end-point of a postmodernist approach (see Gellner 1992: 36–7). However, hardly qualifying as a postmodernist, I persist in the endeavour with the justification that the majority of the books just cited deal exclusively with international migration. Only Parnwell's (1993) introductory text and a major collection edited by the United Nations (1994) attempt to cover both internal and international movements as well as refugees. The United Nations book deals exclusively with the less developed world and leaves aside a consideration of migrants in the principal destination areas. A recent book with the interesting title of *Place, migration and development in the Third World* is essentially restricted to an examination of internal movements in parts of Spanish America (Brown 1990). Thus, there still appears to be scope for an overview of total mobility at the global level.

In addition, the sheer number of publications on migration is taking us to the stage where it is difficult, perhaps impossible, for any single individual to keep track of what is going on. This is leading to problems of awareness and a fracturing of the subject into area-specific and time-specific slots, which may continuously cause researchers to reinvent the wheel. For example, those working on developed countries may not be aware of relevant work carried out in developing countries (see Skeldon 1995b), those specializing in international migration rarely consider the literature on internal movements, and those working in one discipline may not be aware of research carried out in other disciplines. The 'circle of mutual ignorance', to which John Friedmann and his colleagues drew attention over twenty-five years ago is today even more in evidence (cited in Simmons, Díaz-Briquets and Laquian 1977: 69). I do not claim to have any infallible knowledge of this vast and complex field, particularly as I live and work in areas far from the academic heartland of the subject and away from major libraries. Nevertheless, I will attempt to provide a guide through this labyrinthine field and review the main directions of work on migration within what I hope will be a useful, comprehensive and novel framework. While not attempting to create any grand theory that can adequately explain all migrations, this book will try to outline a framework in which internal and international movements can be integrated and their role in development understood.

The structure of this book

A short historiography of writing on migration and development is presented in Chapter 1. The ways in which we have looked at the subject have changed over time, and the main theoretical and ideological approaches are reviewed from the seminal generalizations of Ravenstein in the late nineteenth century to recent postmodernist interpretations. Rather than providing a detailed description of the various models, which are readily available in other books, the emphasis in Chapter 1 is on how our thinking on migration and development has changed, highlighting the contribution that each approach has made to our knowledge of the subject.

No discussion of any topic is possible without a division into analytical categories. This task is especially difficult in the social sciences, where there are few clear boundaries separating the phenomena that we wish to analyse. Globalization and global perspectives are all very well in theory but how, in practice, is it possible to divide the topic into units meaningful for analysis? A general systems approach is advocated in Chapter 2, and the changing global system of migration over the last 150 years is sketched. Within this general framework, three ways of dividing up the global system are proposed: a system of development regions; a system of urban and rural sectors; and systems of different types of migrants. All three are used in this book, either explicitly or implicitly, but the basic framework is a series of migration-development spatial units.

It is argued that we can best conceptualize the relations between migration and development by dividing the world into five development tiers. The term tier is used, first, to give the impression of hierarchy, from more economically developed areas to less economically developed areas and, second, to try to convey the idea that the boundaries between them are fluid. The term 'region' always seems to convey the idea of something either natural or enduring. The tiers are neither natural nor enduring. Chapters 3 to 7 examine the relationships between and within the five development tiers. The first of these is the 'old core' of western Europe, North America and Australasia. The evolution of migration within and between them over the last 500 years is traced in Chapter 3. The historical approach is adopted so that comparisons and contrasts can be noted with the experiences of other areas as they have moved from rural to urban societies and from lower to higher levels of economic development. In Chapter 4, the experience is considered of the major non-European areas in achieving high levels of development, centred in East Asia. The recency and rapidity of this development are clearly major differences from the old core tier, and the changing patterns of migration are examined in this context. The most rapidly growing economies today, and those with a high potential for growth, are considered in Chapter 5, the rapidly expanding core. Most of this tier is contiguous with the core tiers, although there are important outliers of development. In all these areas, the transitions in mobility from one pattern of migration to another are outlined.

Although there is much migration that originates within the above three tiers, these are characterized more as destinations of movement. Similarly, although there are destinations of migration within the remaining two tiers, the labour frontier (Chapter 6) and the resource niche (Chapter 7), these are characterized more as origins of movement. Within this general framework, however, it will be seen that most of the movement is within these tiers, except for the labour frontier, which, by definition, is a source of labour for tiers above and below it in the development hierarchy. Particular attention is paid to the areas along the interface of the tiers to see how the patterns of migration and development are changing there. It will be clear from the discussions that, although there is much talk of a global system, many parts of the world are only tenuously linked to that system and, for some, the linkages are weakening rather than strengthening.

In the Conclusion, I try to tease out the main themes that have emerged in the previous discussions and return to the variation that is to be found in each tier at the local level. In the process of globalization, a succession of local responses is produced. It is how these local variations are linked together that gives form to the global system, with migration providing one of the critical linkages. The book concludes with a review of the major current issues in migration as we move into the twenty-first century.

This book seeks to avoid the standard analyses of explanations for migration: of why people move. There will be little about individual decision-making in this book or even much about the characteristics of individual movers. These are all well covered in existing books on migration. I have striven for a truly global view, to fit all societies into a single framework to examine how they interact, or do not interact, as the case may be. Clearly, a global perspective must, almost by definition, adopt a macrolevel approach. Nevertheless, in places, a microlevel view is required to illustrate particular cases, and the validity of such an approach as a critical component in the analysis is never in question. The combination of regional framework, historical approach and global, comparative perspective offers, I hope, unique insight into the complex and topical issue of migration and development.

Chapter 1

Theories and approaches

The ways in which we have looked at both migration and development have shifted considerably over the years. The approaches
espoused in the literature have changed, with some methods falling
out of favour as new techniques and theories have been introduced.
It would be satisfying to think that these shifts were purely the
result of rigorous scientific investigation but they are as often likely
to reflect changing ideologies, and what are considered acceptable
views at any particular time, as they do any real advance in
knowledge. However, such a view may be too disparaging as the
shifts have made important contributions to development studies
in particular, as well as to work in migration.

Most texts on development draw out the emphasis given to
economic growth in development studies during the early years of
the subject between the end of the Second World War and the
1960s. Then there was a change to a broader concern with social
issues in the basic-needs approach from the early 1970s. More
recent thinking has incorporated environmental issues in 'sustainable development'. Within this broad sequence of change in
emphasis, although not shifting contemporaneously with it, have
been various ideological biases representing those favouring the
free-market approaches of neoclassical economics and those lean-
ing towards a greater degree of state intervention. The individualist
approaches of the free market were contrasted with structural
approaches, which focused on the need to change institutions.
Marxists and neo-Marxists, dependency theorists and those
favouring a Maoist philosophy, and more free-market thinkers have
all made their contribution to development theory at particular
times (for a series of useful charts, see Edwards 1985). One of the
most comprehensive reviews of all these 'competing paradigms'
will be found in Hunt (1989), and the early and pioneering work of
Brookfield (1974) is also useful. No comparable assessment of
migration theory yet exists (although see Massey et al 1994). As
part of population studies, migration has a strong empirical orientation. There is still a marked preoccupation with measurement
and the provision of basic data to the possible detriment of the
development of theory. Nevertheless, migration, as stressed in the
Introduction, is an integral part of development and has been given
considerable attention in the development literature. Readers
interested in the important conceptual issues behind the measure-

ment of migration should pursue these through the various population or demography texts such as, for example, Jones (1990), Bogue (1969) and Courgeau and Lelièvre (1992).

A persistent theme running through migration studies is the issue of analysis at different levels. Migration is the result of the behaviour of individuals, but equally it has an aggregate social form. Migration can thus be analysed at the individual or family level or at the level of the broader social group, depending on what emphasis one gives to the key determining factors. There has been a distinction between those who have given preference to individual decision-making processes as the key to understanding migration on the one hand and structuralists, who favour the analysis of the macrolevel social and economic factors that facilitate and constrain movement on the other. I have argued that the examination of the reasons why individual people move is quite distinct from the examination of the reasons for migration as a group phenomenon (Skeldon 1990: 130). Any emphasis on structural factors, however, leads to the charge that individuals are little more than automatons responding to external stimuli, whose actions are determined by economic and social structures. Migrants are agents whose actions can have consequences, either intended or unintended, upon social structure. The interaction and mutual dependence between individual and structure have been captured in the theory of structuration of Anthony Giddens (1981), which has made a considerable impact in geography and the social sciences in general since the early 1980s. The results of this interaction produce a greater variety of outcome than would be allowed from the single aggregation of individual decision-making. The tension between structure and individual will underlie much of the discussion in this book but will be implicitly accepted rather than explicitly addressed. The increasing emphasis on variety of outcome, as raised in the Introduction, has also introduced to migration studies the latest ideological wave to affect the social sciences, postmodernism. Its reaction against scientific reasoning and unilinear 'meta-narrative' has brought a very different perspective to migration research that will be touched upon later in this chapter and again in the Conclusion.

The present chapter will focus on the major trends in migration research specifically related to development issues in a brief historiography of the subject. We begin with its antecedents in the work of Ravenstein and move on quickly to the main trends in the literature after the Second World War.

Antecedents to models of migration and development

In the work of the founding father of modern migration research and analysis, E. G. Ravenstein (1885, 1889), it was implicit that migration was in effect caused by economic development. Two of his famous 'laws of migration' make the relationship clear: that 'migration increases in volume as industries and commerce develop and transport improves' and that 'the major causes of migration are economic'. Migration from Ravenstein's point of view thus appeared to be a consequence of development. It also symbolized development. 'Migration means life and progress, a sedentary population stagnation' (Ravenstein 1889: 288). A total of eleven laws, principles or rules of migration can be identified from the various writings of Ravenstein and these can stated as follows (from Grigg 1977):

1. The majority of migrants go only a short distance.
2. Migration proceeds step by step.
3. Migrants going long distances generally go by preference to one of the great centres of commerce or industry.
4. Each current of migration produces a compensating counter current.
5. The natives of towns are less migratory than those of rural areas.
6. Females are more migratory than males within the kingdom of their birth, but males more frequently venture beyond.
7. Most migrants are adults: families rarely migrate out of their county of birth.
8. Large towns grow more by migration than by natural increase.
9. Migration increases in volume as industries and commerce develop and transport improves.
10. The major direction of migration is from the agricultural areas to the centres of industry and commerce.
11. The major causes of migration are economic.

These generalizations (few social scientists today favour such a definitive term as 'law') were derived from an empirical analysis of the data for nineteenth-century England, and Grigg (1977) showed that many of these were still valid in the context of migration in Britain today and that several others were still worth testing. While this is not a major theme of this book, we can nevertheless bear the generalizations in mind during the discussions of migration to see the seminal importance of Ravenstein's work as several of his generalizations are relevant not only to internal migration but also to international movements.

The economic basis for movement remained in many subsequent interpretations of migration, even when social variables were also introduced. These could all be categorized as 'development' variables of one type or another and they were explicit in the attempt by Everett Lee (1966), over eighty years after Ravenstein, to elaborate a general 'theory' of migration which would provide a schema of the factors that could explain the volume of migration between any two places. This attempt was essentially a descriptive model of migration incorporating a series of 'pushes' from areas of origin and 'pulls' to areas of destination. These pushes and pulls tended to be couched in a dualistic way: the pushes from origin were the polar extremes of the pulls to destinations. Lack of job opportunities in villages, as opposed to the existence of job opportunities in towns, was seen to cause rural-to-urban migration, for example. Other means of earning income such as access to land, and more social factors such as education, health and housing, were similarly incorporated into this descriptive model of migration. Rural areas were also seen to have their pulls (community life, relaxed lifestyle, for example) and urban areas had their pushes (congestion, crime and so on), but these models were almost entirely descriptive and rarely attempted to relate, in any rigorous way, how the development variables influenced migration.

The pushes and pulls leading to migration were generally seen to be created by two main forces: population growth in the rural sector that brought a Malthusian pressure on agricultural resources and pushed people out, and economic conditions generated mainly by external forces that drew people into cities. As Williamson (1988: 426–7) has argued in an important review article, demographers and early development economists favoured the former interpretation, while most economists by the 1980s had turned towards the latter interpretation.

Among the approaches which adopted population as a key driving force were a family of models that arguably have been the most influential in relating migration and development in recent years. These evolved from work in economics which saw the migration of people from a rural labour-surplus economy to an urban labour-deficit economy as an essential component of the whole development process. These were the neoclassical dual-sector economic models of migration. They stemmed from the work in the 1950s of the Nobel laureate Arthur Lewis and saw migration as the 'natural process' (Todaro 1994: 260) by which surplus labour in the rural sector would be released to provide the workforce for the modern urban industrial economy. Migration in this sense was necessary for development. The subsequent experience of developing countries belied this optimistic interpretation as the industrial sector did not generate sufficient jobs to absorb the migrants from the countryside. Instead of migration slowing in the face of increasing urban unemployment, as might be expected from the assumptions of the equilibrium model, it in fact was seen to persist and even to increase.

In the late 1960s, Michael Todaro outlined the basis of a very influential model that attempted to explain the apparent anomaly between rising urban unemployment but continued high rates of rural-to-urban migration (Todaro 1969, 1976). This model was based on expected rather than real income differences between the rural and urban sectors. Potential migrants as individual decision-makers would 'consider the various market opportunities available to them as between say, the rural and urban sectors, and choose the one which maximised their "expected" gains from migration' (Todaro 1976: 28–9). A potential migrant could thus discount periods of unemployment against expected higher income, once access had been gained to an urban job. Over the longer term, this strategy would yield more gains than continuous but low-paid rural occupations.

The strength of this approach, quite apart from its analytical simplicity, was that it focused attention on the rural sector and on rural development. The model showed that increases in the number of urban jobs or urban incomes would lead to still further increases in rural-to-urban migration rather than alleviating the urban unemployment problem. Rural dwellers responded to these increases by moving to towns in ever larger numbers in the expectation of finding one of these new jobs sooner or later. Hence, attempts to alleviate urban unemployment simply by creating more urban jobs were likely to be self-defeating as they merely raised expectations and accelerated migration. Measures to solve urban unemployment had therefore to be taken as much in the rural sector, to improve conditions there, as in the urban sector. The principal value of this approach was that it drew attention to the linkages between rural and urban sectors and to the centrality of migration in any programme of integrated development. The model spawned a whole series of studies of internal migration in developing countries which emphasized the critical role of population movement in development.

> We must recognize at the outset, therefore, that migration in excess of job opportunities is both a symptom of and a contributor to Third World underdevelopment. Understanding the causes, determinants, and consequences of internal rural–urban labor migration is thus central to understanding the nature and character of the development process and to formulating policies to influence this process in socially desirable ways.
>
> (Todaro 1994: 241–2)

The weaknesses of this whole approach are now well known and revolve around its assumption that migrants are individual actors who are seeking to maximize income in undifferentiated labour markets.

New economics approaches

An understanding of the causes and determinants of migration is more likely to be achieved by conceptualizing migrants as being embedded in tight social and economic networks and attempting to minimize risk within highly segmented labour markets. Rural populations do not have access to perfect or comprehensive information about urban job opportunities; they have information that is selective and that comes through particular channels. Most migrants arriving in the city have a very good idea of what is lying in store for them but their knowledge may encompass only a single community and perhaps one occupation. The concept of migrants freely competing in an integrated general labour market is false.

Migration can be as much a family or group decision which seeks to exploit opportunities in various 'niches' spread among origin and destination areas as it is one for individual improvement. Resource diversification, rather than income maximization, thus becomes a central explanatory variable of migration (Stark 1991). The city, or the plantation, or any sector in the non-traditional rural sector, in effect becomes part of the resource base of the peasant household to be 'harvested' regularly (permanently or seasonally) or intermittently, along with village lands, as part of a survival strategy (Hugo 1994b). This type of approach can be broadly seen to fall under the 'new economics of migration' category of explanation to distinguish it from the previous neoclassical models to the extent that income-maximizing strategies have been replaced by risk-minimizing strategies. The distinction between the two approaches has already been drawn for studies in international migration (Massey et al 1993, 1994). However, the approach as described here would also incorporate two other theories that Massey and his colleagues identified separately for international migration, namely, segmented labour market theory and network theory. Risk minimization can only be implemented through networks, and the networks established by families serve not only to perpetuate the migratory flows but also to control access to particular labour markets. These issues will become apparent in the more detailed discussions of migration in particular areas in subsequent chapters and have also been treated at length in a previous work (Skeldon 1990).

While risk-minimization strategies within family networks as an explanation for migration are still very much centre stage in current approaches to the topic, an important and serious challenge has been raised by those analysing gender relations (for example, Hondagneu-Sotelo 1994). The unity of an integrated family acting to access different areas may be more presumed than real, and there can be significant differences of opinion within the family, with wives and daughters migrating despite the wishes of husbands and fathers. There are female networks, separate from those of the men, which influence decisions to move. It is unlikely that there will ever be a retreat back to the primacy of the individual decision-maker in a social vacuum, but one of the challenges of future research will be to balance gender and generational networks within an integrated structured model of migration.

The neoclassical and new economics approaches to the study of migration both highlighted its role in the transformation of traditional rural societies consequent upon the movement to new centres of influence, or 'modernization', primarily towns and cities. Studies of internal rural-to-urban migration in the developing countries have certainly continued apace, but the emphasis in the consideration of issues of migration and development has broadened over the last ten to fifteen years to place international migration firmly at the centre of migration studies. This trend has evolved to such an extent that internal migration has been reclassified by some as 'population redistribution' (see, for example, United Nations 1994), which is usually associated with urbanization. Migration has come to mean only international movements. The *Age of migration* of Castles and Miller (1993), for example, is an age of international movements just as Myron Weiner's *Global migration crisis* (1995) is a crisis of growing international movements. Clearly, international migration is not a product of the last ten to fifteen years and it was of concern long before the recent upsurge in interest. However, previously, the interest focused on movements within what could be considered the developed countries of the then First World and primarily those of the Atlantic Community (see, for example, Thomas (1954)). Now the concern is with the movement out of poorer countries towards the richer countries of Europe and North America and among the poorer, developing countries themselves. Yet the number of internal migrants is currently many times that of international movers and the latter represent, as we shall see in subsequent chapters, not an exceptionally large proportion of their origin societies by historical standards. The reason why international migrations have moved to the centre of concern in the overall field of migration studies is essentially political. The current increase in numbers represents a challenge to states in many parts of the world, not just in the most developed regions, and a perceived threat to their sovereignty and cultural integrity. Migration, essentially international migration, has thus become an increasingly important part of concepts associated with international relations.

Many of the conceptual approaches used to study international migration have been the same as those used for the study of internal movements. For example, the neoclassical and the new economics approaches have again been prevalent as have segmented and dual labour market theory and network theory (Massey et al 1993, 1994). As mentioned above, however, the differences between the latter three approaches in internal migration may not be as clear-cut as Massey and his colleagues argue in the case of studies in international migration. Approaches which have been much more prevalent in the study of international rather than internal migration have been those based upon a world systems perspective. These are associated with the expansion of international capitalism and the trends towards a global economy. While internal migration can, although mistakenly I would argue, be seen in isolation and restricted to one country, international migrations certainly cannot. The perspectives adopted must be much broader and the concepts of the world system and globalization, with their emphases on linkages and striving to understand the totality, are singularly appropriate to the task. The approaches discussed thus far tend to have evolved out of economics or the economics of development, while the world systems and global approaches have come more from sociology and from the political science of international relations.

World systems and global approaches

The world systems approach originally espoused by Immanuel Wallerstein (1974, 1980, 1989) traced the evolution of a global economy through the expansion of Europe from the end of the fifteenth century. It envisaged the creation of an increasingly interdependent economic system as European capitalists gradually sought out resources throughout the world. These resources in the periphery were required for the development of the core. Whether the primary driving forces were indeed economic while ignoring political factors (Zolberg 1981) need not detain us here. What is important is the paradigm shift of many branches of the social sciences towards a global framework to such an extent that it is difficult to envisage any analysis in urban studies, development or population that does not have at least some global dimension (see, for example, King 1990: 3–4).

Human migration was an integral part of the creation of the world system through the establishment of European colonies and trading-posts across the globe which linked the capitalist system through an expanding and interacting human network. The process of decolonization and the increasing number of independent nation states after 1945 did not lead to a decline in globalization. Certainly, in a few cases it did, as certain countries attempted to break out of the system and establish their economic and political independence through systems of self-reliance or in regional systems separate from the capitalist system. These cases most specifically applied to the socialist economies of the then Second World and countries in the Third World within their ambit. The collapse of the Soviet system in 1989 and the opening-up of China since 1979 after thirty years of isolation have lent support to the idea that globalization is leading to a more homogeneous world. The migration of peoples within global networks of circulation requires common means of communication and diffuses common values and ways of doing things. Yet, as a subsequent section of this chapter will show, there are other outcomes. Although there may be an overall loss of diversity – perhaps 50 per cent of the world's languages will disappear by the end of the century, for example (*Geographical Magazine*, May 1995: 8) – there is certainly not going to emerge, in the short term at least, one single political or cultural system that signifies an end of history or the attainment of some social and political nirvana. There is also the spread of diversification as well as homogenization, and processes of localization as well as globalization.

While migration is a proximate cause of the development of global linkages, one of the principal driving forces is the transnational corporation. While most of their activity is concentrated in the more developed parts of the world, transnational corporations are instrumental in extending capital into the periphery in the search for new markets and to exploit resources. Their global reach has created the functional interdependence of regions across space that has become known as the 'new international division of labour', in which each region capitalizes on the part of the production process in which it has the greatest comparative advantage, and regions of specialization emerge. For example, and at the simplest possible level, those manufacturing activities that are labour-intensive diffuse to those areas where there are abundant supplies of cheap labour, while high-cost labour areas will support mainly capital-intensive manufacturing and high-level service activities. That there are deviations from this simple pattern does not deny its essential logic, and the spatial divisions of a global economic system will be used as a basic framework in this book. These spatial divisions will be discussed in Chapter 2. The other major actors in the creation of a global economic system are the international banks and the nation states (Cohen 1981). The tension between the state, which sees its national authority challenged, and the transnational banks and corporations, which are responsible to no government, creates much of the turbulence and constant change that characterize the present economic system. For an excellent review of the role of the transnational corporation and the globalization of the economy, see Dicken (1992).

The flows of goods, capital and people articulate the global economy. Yet, although the numbers of migrants crossing international borders have unquestionably been increasing and there has been an internationalization of labour markets, the movement of people does not appear to have matched the flows of goods or capital. 'It is the relative immobility of labour that is the significant backdrop to growing economic integration' and in Europe, where there is free movement across boundaries within the Community, only 1.5 per cent of the labour force are to be found outside their countries of origin (Campbell 1994: 186–7). Clearly, this economic approach is a very different point of view from the more politically and sociologically oriented age of migration and migration crisis interpretations discussed earlier. The immobility thesis may, however, be overstated as there are many other forms of short-term labour mobility that do not show up in migration statistics, and no account of internal migrations was taken into consideration. Labour may not move as far or as fast as capital, but it moves nevertheless and is becoming an increasingly important component of the global system.

Perhaps the major weaknesses with the world systems approach are its Eurocentric bias and its implication that development in the periphery is dependent mainly upon its linkage to the core. Many of these global approaches thus share the legacy of modernization theory with its assumption of the diffusion of 'modern' influence out from (western) centres to galvanize indigenous societies along some road to progress. The opposite side of the coin was that the linkages, rather than developing the societies being contacted, would lock them into positions of perpetual powerlessness and dependence upon the west. This was the basis of dependency theory and both it and the earlier modernization theory are well established in the tradition of development studies (see Todaro 1994). Both assume that the societies being contacted are relatively passive and ineffective in their dealings with the west. We know that this is quite incorrect, with western nations reaching a state of technological superiority only in the nineteenth century, and that there were important states in Asia and Africa which rivalled and resisted European influence in their regions as they too extended their control over populations and labour (see Blaut

1993; also Adas 1989, Reid 1988). Nevertheless, the world systems and global systems approaches are important from a conceptual point of view in that they can provide a useful framework for the examination of migration because of their emphases upon functional economic linkages across space. Additional concepts, however, are required to provide a more balanced weighting for origin and destination areas. The communities established through these global movements have given rise to a term which is becoming an increasing part of migration studies and which introduces different concepts and new ways of looking at international migration. That term is 'diaspora'.

Diaspora

Diaspora, from the Greek 'to scatter' (sow), has been primarily associated with the dispersal of the Jews out of Israel, although it has been applied to any people living outside their traditional homeland (OED). An ancient word, its use in English was fairly limited until the 1990s, when there was a sudden explosion of interest with the appearance of a new journal, *Diaspora*, a series of books promised on the topic, edited by Robin Cohen, as well as many individual books and articles, including *The Penguin atlas of diasporas* (Chaliand and Rageau 1995), dealing with particular aspects of the subject. Diaspora made a somewhat earlier appearance in French, and much of the conceptual debate has been carried out by the French, but here again it is a relatively new concept grafted on to an old word. Its first broader application was to trading diaspora (Cohen 1971, Curtin 1984) to convey the idea of a transnational network of trading communities interacting socially almost entirely within that network and having relatively few relations, apart from the strictly commercial, with the peoples among whom they lived.

Obviously, all diasporas involve a migration but not all migrations make up a diaspora. There is no difficulty in associating the Jews or the Armenians with the term but it is almost never applied to the British (although see Kotkin 1993) and its frequent application to the Chinese and the Indians is problematic. According to Sanguin (1994), diaspora has had both highly general and highly specific applications. In general uses, diaspora has referred to the movements of any distinct ethnic group to any other part of the world, while the specific reference restricts the usage only to movements that have been massive and where the numbers who have left their homeland greatly outweigh those that remain. While the former usage could apply to virtually any migration, the latter could refer only to the movements of such peoples as the Jews, Armenians, Lebanese, Palestinians and Irish. Although what of the Scots, whose descendants overseas outnumber those remaining at home and who religiously maintain Scottish traditions in places far from Glasgow or Aberdeen? Sanguin (1994: 495) himself suggests that diaspora could have a middle-range usage applying to those ethnic communities which have come from countries where there is still misery, overpopulation, insecurity, dictatorship or religious or racial discrimination. This seems to fit the application of semi-diaspora by Chaliand and Rageau to the Chinese and Indians, but such qualification of terms does not facilitate precision. See the review of Chaliand and Rageau on the Internet by John Radzilowski in February 1996 (H.Ethnic@msu.edu).

The movements that can thus be classified as 'diasporic' under these categories will be many. However, there appears to be a common thread of compulsion in creating a diaspora. People are being expelled or pushed out of their homeland in diaspora and perhaps one of the reasons that the term is becoming such an important theme in current migration work is the increasing number of refugees or asylum-seekers in the world. Just as internal and international migration were so long separate fields of enquiry, so too were refugees seen as being quite distinct from other categories of movers. Redefinition of some refugees as 'economic migrants', as in the case of many of the 'boat people' who left Viet Nam, and the realization that economic and environmental conditions in the home areas, as well as political persecution, could render those places intolerable blurred the distinction between refugees and other types of migrants and emphasized that forced movements needed to be considered with other forms of migration within a unified framework. Theory-building to incorporate refugees and other types of migration within systems of political development is, however, still at a fairly early stage (see Kunz 1981, and Skeldon 1990 for a brief review).

Hence, diaspora appears to emphasize a compulsory, non-voluntary component in much modern migration, but a flight from poverty and deprivation as much as from the threat of persecution or actual violence. Diaspora draws attention to the migrant as victim, not only in the homeland but also at the destination, where minority communities are established. In destination areas, isolation, separate identity and discrimination are more recurrent themes than integration or assimilation. This has profound implications for the concepts of nationalism and nation-building in destination societies as diaspora emphasizes roots, exile and home. It highlights links back to the origin rather than forward to the destination, identifying specific migrant groups rather than states as key actors. While the migrant as victim is still central to the diaspora approach, this view needs to be modified by the very real and positive contributions that migrants can make through their exile (Cohen 1996a). The theme of dynamic migrant groups taking skills around the world to the benefit of host societies as well as of the migrants themselves, and independent of the reasons why the migrants might have left in the first place, is central to Sowell's (1996) interpretation. Irrespective of whether the migrants are victims or pioneers, central to the diaspora approach is the focus on networks: within the communities of destination, within those organizations that maintain group identity, and between origin and destination areas. Nation becomes associated with global migrant cultural groups rather than origin or destination states. Migration rather than a move for settlement is, from the point of view of diasporas, a transnational system of circulation.

Just as those studies of mobility took so long to make an impact on studies of internal migration, so diaspora has only recently become one of the leading themes in international migration research. It appears to be associated with a resurgence of interest in culture and in cultural studies: of the different identities of migrants rather than the unity of a new blended destination culture. The destination areas may, however, create new 'cultures of exile' reproducing neither the culture of origin nor a synthetic common destination culture but an idealized concept of home culture which is a kind of ethereal 'between' culture, rootless and rejected by both origin and destination. These cultural approaches to migration and diaspora have been given prominence in the recently (1991) established journal *Diaspora* and are part of post-modern interpretations of society and culture discussed later in this chapter.

While some approaches to diaspora draw attention to the 'between' nature of migrant groups and exiles, the concept itself shifts attention away from the migrant as being uprooted, as in Handlin's (1951) classic study, and divorced from his or her origins. The migrant is moving in networks embedded within cultural systems which are transferred to new situations. Continuity rather than discontinuity, and multiculturalism or polyethnicity rather than assimilation, are the themes which emerge through the application of diaspora approaches.

Whether diaspora communities will participate fully and con-structively in the life of the host nation and how these communities relate to the origin communities become key concerns. Migrants are almost always minorities in their host communities, in both internal and international migration, and how they participate in the economy and society of destination areas is obviously of great importance to their hosts. If they come from very different types of societies ideologically and maintain close relations with those societies, their presence may be viewed with suspicion by their hosts. On the other hand, if they come from minority groups within the origin areas, their continued presence in another area may be seen as a threat to origin governments. In some cases, the migrants may seek to destabilize the government in their home areas and, in so doing, create problems within the international system between host and origin governments. Or destabilization may be officially sanctioned by the host government if there is open or latent confrontation between governments. In all these cases, migrants can pose important questions of security to either host or origin countries. These aspects are given detailed consideration in Weiner (1993).

Even without these more overt aspects of security, migrants transfer ideas and technologies from one area to another that can have a significant impact on development. The removal of German scientists and engineers to the United States and to the Soviet Union after the Second World War had great significance for the development of later missile and space programmes. The migration of Islamic peoples into core cities of the developed world has provided safe havens for a variety of groups which may be unsympathetic to the goals of western secular governments. The return of peoples with long experience in the developed world or in exile can have a profound impact on the direction of political and economic development of home countries as the contrasting examples of returned students from the United States to Taiwan or South Korea on the one hand, or of the Ayatollah Khomeini from France to Iran, on the other, can demonstrate.

Diaspora also focuses attention on the origin countries and cultures as expansionary groups, either being pushed out or moving out to exploit new opportunities in what Kotkin (1993) has called 'the making of global tribes'. Kotkin's use of diaspora clearly falls into Sanguin's 'most general' of applications as he gives attention to the British diaspora as the 'largest cultural and economic diaspora in world history' (Kotkin 1993: 22) and essentially laying the basis for the modern world. Such a broad interpretation of diaspora may be such that it is rendered virtually meaningless. Nevertheless, the use of the term 'diaspora' does draw attention to linkages between origin and destination. Incorporating the expansion of particular groups not only emphasizes those linkages between a core 'home' group and peripheral 'away' group but also raises critical issues of identity and political participation and this gels with the global economy models of development previously discussed.

The emphasis on linkages through the diaspora interpretation brings into question several stereotypes and none greater than the settler/sojourner contrast between European and Asian migrants to

North America in the latter part of the nineteenth century. The Europeans went as settlers to create a new culture in the melting-pot of the United States, while the Asians went as sojourners with the express intention of returning. Europeans were not in diaspora whereas Asians, and particularly the Chinese, were. It is now clear that many Europeans also went either with the intention of returning after a season or a short stay, or with the intention of returning home after some time in North or South America (see Chapter 3). As we will see in subsequent chapters, there are definite parallels and the settler/sojourner contrast does not survive close scrutiny.

Towards the end of their review of theories of international migration, Massey and his colleagues (1993: 454) observe:

> The various propositions of world systems theory, network theory, institutional theory, and the theory of cumulative causation all suggest that migration flows acquire a measure of stability and structure over space and time, allowing for the identification of stable international migration systems.

They further argue that migration systems theory – first developed by Akin Mabogunje (1970), although not referred to by Massey, presumably because that seminal article referred to internal movement – provides a useful framework to give substance to these spatial and temporal structures. The advantage of a systems approach is that it draws out the feedback role of migration between origin and destination to show how both are transformed economically and socially through the linkage of migration (Mabogunje 1970, Kritz, Lim and Zlotnik 1992). Systems can indeed be identified for migration, and much more will be said on this in Chapter 2, but an even more general approach not referred to at all by Massey and his colleagues which could sharpen the analysis of the spatial and temporal structures is what might be called 'transition theory'. Like systems theory, it is not a theory as such, but a series of generalizations that can help to provide an order for the analysis of the complexity of the real situation.

Transition theory

Over the last 250 years we have seen a transition in the developed countries of Europe from a primarily rural society to a primarily urban society (the urban transition), from societies of high fertility and mortality to societies of low fertility and mortality (the demographic transition), and from economies based upon agriculture through those based upon industry to those based upon services (a development transition?). Into this came the mobility transition proposed by Wilbur Zelinsky, which, in a very general way, attempted to link the demographic transitions and modernization with changes in the type and pattern of migration:

> There are definite patterned regularities in the growth of personal mobility through space-time during recent history, and these regularities comprise an essential component of the modernization process. (Zelinsky 1971: 221–2)

The strength of this approach is that it combines within a single framework several different types of population movement, including both internal and international migration. The former was further divided into movement towards the agricultural frontier, from rural areas to urban areas, and within and between urban areas themselves. Zelinsky recognized that permanent, or at least long-term, migration represented only one type of population movement and that there were a whole series of short-term forms that he classified under circulation. These five mobility types, he hypothesized, varied in relative importance over time but they could generally be correlated with the stages of the demographic transition. These are shown in his classic diagrammatic form (Fig. 1.1), which, like the demographic transition, roughly described the experience of western European countries.

While the nature of the relationship between modernization, which can be assumed to represent development, and the form of mobility was never precisely stipulated, part of the innovative nature of the approach was to identify two clear connections between technological change and mobility change, and these are shown in the last two panels of the diagram. As forms of transport improved, people could stay in their places of origin and commute to jobs, whereas before they would have had to move to these jobs. Thus, over time, some forms of migration are absorbed by circulation. In addition, as forms of electronic communication improved, people would be able to work from home rather than commuting to an office, so that, over time, circulation could be replaced by increased usage of the electronic media.

The weaknesses of the approach have been widely discussed. First, there are errors of fact. The model perpetuated the myth of the immobile pre-modern society. As stressed in the Introduction, contemporary studies in many parts of the developing world have emphasized the importance of non-permanent forms of mobility or circulation, and so too have studies of pre-modern European societies. The idea that the Industrial Revolution uprooted peasants from their villages for the first time was more a romanticized elitist view of peasant life than an interpretation based on objective analyses. Recurrent disease, famine, warfare, slaving expeditions, pilgrimages, as well as state-sponsored colonization, ensured almost continuous population movements, albeit with distinct fluctuations in volume throughout human history. Hence, circulation, as depicted in graph E of the diagram, requires major modification. Elsewhere, I have attempted to reconstruct a mobility transition that might better 'fit' the historical record (Skeldon 1990).

The second weakness is that the intent of the mobility hypothesis to relate mobility change to the stages of the demographic transition was never realized. Zelinsky simply associated changing forms of mobility with five phases of fertility and mortality change that marked the transition from high fertility and mortality through

Fig. 1.1 The changing importance of various types of migration in a mobility transition.

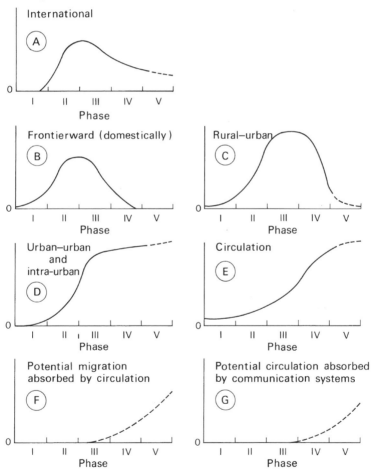

Source: W. Zelinsky (1971) The hypothesis of the mobility transition. *Geographical Review* **61**(2): 233.

to low fertility and mortality. Just how mobility might affect fertility or mortality remained unanswered. Clearly, long-term circulation leading to the separation of spouses is likely to depress fertility (Potter and Kobrin 1982). The movement of migrants from areas of high fertility and mortality to areas of lower fertility and mortality may be marked by the migrants assuming the fertility and mortality characteristics of the area to which they move. However, migrants from high-fertility areas may be positively selected and have lower fertility than the norm for their region of origin and hence there may be little 'migration effect' on fertility. Conversely, migrants from high-fertility areas may indeed transfer their high fertility to the destinations.

A key issue is whether the circulation or return movement back to their home areas of migrants with lower fertility behaviour influences the rest of the population in these areas. It might appear intuitively obvious that migrants may be key agents in the diffusion of new ideas of family size or in ideas that will ultimately affect fertility and mortality, but the precise nature of these relationships is not so obvious. Davis (1963) included migration as part of his 'multiphasic response' to increasing rates of population growth in which migration from rural areas, or a country, is functionally equivalent to the postponement of marriage, increased celibacy or the adoption of contraception. Tests of these ideas were carried out in several countries at various levels of development to support the idea that both rural-to-urban migration and emigration could substitute for fertility decline and slow the rate of population growth (Mosher 1980a,b, Kim 1994). Despite this work on population growth and migration within the context of the model of the demographic transition, the actual relationship, if any, between migration and fertility decline remains vague. The examination of changing patterns of migration in the context of the other two demographic variables remains a critical and still largely ignored area which could benefit from further research (Skeldon 1990, 1992b).

The most obvious potential impact of changing fertility and mortality levels also remains unexamined in the mobility transition: cohort effects. The changing balance of births and deaths through the demographic transition alters the age structure with, very generally, an increasing youthfulness in the early phases of the transition and an ageing of the population in the later phases. As most migrants tend to be young adults, the absolute number and proportion of this sector of the population is going to be a factor in changing patterns of mobility. Cohort effects are unlikely to be the major factor accounting for shifts in the pattern of human movements, but they are likely to be one factor. An interesting variant of transition theory and cohort effects has been made in the study of ageing populations. The 'elderly mobility transition' states that, as populations age through the transition to low fertility, the rates of migration of the elderly increase. This idea has served as a useful framework for the study of mobility in ageing populations (Rogers 1992).

While the demographic transition has proved a very useful general descriptive model against which to examine the experience of particular countries, it is clear that there is not one single path, but that there are many paths through the transition to low fertility and mortality. Herein lies the third, and arguably the most important, weakness of the mobility transition: its depiction of migration and development as a unilinear process that affects all areas in the same way. It is hardly fair to criticize an approach for being a child of its time, as all writing and argument (the current term would be 'discourse') reflect prevailing attitudes, approaches and sentiment. Almost at the same time as Zelinsky produced his mobility transition, Omran (1971) outlined his epidemiological transition, which attempted to show how the pattern of disease and of death changed over time as societies developed, and Friedmann and Wulff argued for their urban transition (1975). The epidemiological transition was later elaborated into the idea of a 'health transition', in which societies are gauged to be moving from an era of infectious diseases towards a stage where the degenerative diseases of ageing societies become dominant. This approach perhaps finds its clearest expression in the work of Jack Caldwell and his associates in the journal *Health Transition Review*. The urban transition was resurrected in a book of the same title some fifteen years after the Friedmann and Wulff formulation (Ginsburg 1990).

These approaches are associated with modernization theory, which depicted societies moving in a steady progression towards ever higher levels of prosperity through the gradual acceptance of western-style technologies and institutions. The development experience of the last thirty years has thrown into question the validity of this approach. On the one hand, the unequal nature of the exchange between more developed and less developed countries has not allowed the diffusion of the necessary infrastructure and institutions that permit countries to develop in the ways most appropriate for them. On the other hand, the internal conditions have often not favoured the reception or adoption of ways and means that might improve conditions within the country. It is clear, however, that there is no single, simple pattern of development and hence no single, simple pattern of mobility change. In our postmodernist world, a 'meta-narrative' such as transition theory has to be modified by local context and experience. Variation, and the reasons for that variation, as stressed in the Introduction, have to be an integral part of any global view that, almost by definition, has to be based on the comparative method. The study of variation must, nevertheless, be tempered by some framework that avoids the seductive traps of exceptionalism and total relativism.

The idea of transition theory is still strong in migration and development studies. Quite independent of the work of Zelinsky – at least the proponents of these approaches make no reference to the mobility transition – two other transitions have emerged in the context of studies on international migration. The first is a four-stage model that describes the migration from the Mediterranean countries to western Europe and to Australia, and from Latin America and Asia to North America. This model provided a framework for the assessment of international migration by Castles and Miller (1993), although it was first developed for western Europe over twenty years before by Roger Böhning (see Lemon 1980), another example of the stage thinking of the time. The four stages in the development of an international migration system were postulated as follows:

> Stage 1: temporary labour migration of young workers, remittance of earnings and continued orientation to the homeland;
> Stage 2: prolonging of stay and the development of social networks based on kinship or common area of origin and the need for mutual help in the new environment;
> Stage 3: family reunion, growing consciousness of long-term settlement, increasing orientation towards the receiving country, and emergence of ethnic communities with their own institutions (associations, shops, cafés, agencies, professions);
> Stage 4: permanent settlement which, depending on the policies of the government and the behaviour of the population of the receiving country, leads either to secure legal status and eventual citizenship, or to political exclusion, socioeconomic marginalisation and the formation of permanent ethnic minorities.   (Castles and Miller 1993: 25)

These stages are purely descriptive and although one stage appears to evolve seamlessly into the other, driven by an intrinsic momentum in the migration process based upon human networks and chain migration, with migration leading to further migration, there is no attempt within the model to link the sequence either with ongoing demographic changes or with economic factors. The implicit economic base is the demand for labour in core developed countries which instigates the process, after which it takes on a momentum of its own. While economic cycles, or direct policy intervention, may vary the length of time that it takes for one stage to develop into another, neither the hidden hand of the market nor immigration policy appear to be able, according to the proponents, to stop the process or cause it to deviate significantly once it has begun.

The second transition in recent work on international migration is the concept of the 'migration transition' (Fields 1994, Abella 1994b). Rather than a global model of the evolution of migration, it seeks to identify those factors which bring about the transformation of labour-surplus economies of emigration to labour-deficit economies of immigration. Much of the attention has been directed towards the rapidly developing economies of East and Southeast Asia. In this approach, the interrelationship between the direction of migration and economic development is explicit, and demographic factors controlling the size and rate of growth of the indigenous labour force are an integral part of the analysis. However, numbers are but one factor and the quality of the labour force in terms of its education and skills training are as, if not more, important. This approach also places migration much more into the international division of labour and the trend towards a global economy. The skills demanded in each part of that global system and how these change with the global shift (Dicken 1992) of industries out of core developed countries are critical elements in this migration transition. More limited in time and space than the mobility transition, it is much more integrated with developmental variables and will be discussed at greater length in Chapter 4.

Lastly, and again in apparent complete isolation from Zelinsky's ideas, there is the attempt by the French demographer, Chesnais (1992: 153–89), to incorporate international migration into the demographic transition in Europe in yet another 'migration transition'. Peaks in emigration from European countries coincide 'within a margin of a few years' (Chesnais 1992: 165), with peaks in natural increase demonstrating the significant cohort effects on migration. Migration, in the view of Chesnais, is a safety valve for excess population and he observed a general progression from west to east across Europe. Countries in the less developed world today do not have access to 'free land' for colonization during their period of most rapid growth and thus follow a different path through the transition. Internal movements are excluded from this transition, as is the importance of return migrations, but Chesnais's conceptualization is one of the few to integrate migration into a model of the changing relationship between population and development. Chesnais (1992: 153) observed that the absence of migratory movements from the theory of the demographic transition was 'curious'. Just as curious is the fact that the few who have been making the attempt at such an integration over the last twenty-five years appear to have been working in isolation one from the other. Nevertheless, despite these limitations and despite the objections directed at transition theory itself, that theory has served, and continues to serve, as an adaptable and valuable framework for the investigation of population mobility.

The postmodernist view

The last and most recent conceptual approaches to migration are the very antithesis of theory and generalization: those literary, cultural and postmodernist views of population movement. In common with so much of the genre, the arguments are rendered imprecise by linguistic pretension – 'metatwaddle', in the telling words of Ernest Gellner (1992: 41). Yet, there are indeed important insights in the best of the accounts which provide a new perspective upon migration, migrants and development. These approaches revolve around the migrant as an individual but not as the individual decision-maker of the behavioural approach of the late 1960s, as in the pioneering work of Wolpert (1965), for example. In the postmodern approach, it is the experience of migration that is important: the experience of moving from one area, from one culture, to another is fundamental for the creation of new cultures. Exile, identity and experience become key concepts. In a post-colonial world, the emphasis has shifted from national concerns towards more universal experience: travel and living overseas become central to the new writing. 'Cultural expatriation is now widely regarded as intrinsic to the end-of-the-century postcolonial literary experience' (Boehmer 1995: 223). For a review and discussion of migration and literature, see King, Connell and White (1995) and for the idea of migrancy in literature, see the critique by Ahmad (1992).

The break-up of colonial empires and the emergence of new nations has also created ever larger numbers of people who have been forced to move into some kind of exile. Yet, despite the hardships and cruelties of the move, the tensions created by being thrust into a new environment have been a critical part of the development of the new intellectual. The experience of migration, of living in two or more places, provides a much deeper insight into the postcolonial condition. One of the paradoxical gifts of imperialism, according to Said (1993: 407–8), was that it created a global interdependent community while at the same time allowing people to believe that they belonged to only one part of that community as whites, or blacks, as orientals, or westerners. The experience of exile, of migration, erodes these simple created identities and makes new ones that are more consistent with the post-colonial world. Migration is essential for the development of mixed cultures and in this the intellectual, often in the vanguard of the migratory flow and amongst the first into exile, plays a key role. The analysis of literature written by the expatriate in exile has provided a novel and vibrant perspective on the whole issue of migration and development which will become clearer in the discussion of students and elite migrants in later chapters. Feelings about place are very much part of the 'new' cultural geography, and homesickness or fascination with the new environment are part of the experience of exile. Yet, to conclude this section with the words of a twelfth-century monk, cited by Said, the culmination of individual development is perhaps the denial of place:

> The person who finds his homeland sweet is still a tender beginner; he to whom every soil is as his native one is already strong; but he is perfect to whom the entire world is as a foreign place.
>
> (Hugo of St Victor, cited in Said 1993: 407)

Discussion

This chapter has sought to outline the main conceptual directions taken in migration and development as espoused in the recent literature. These directions are numerous and their uses have varied over time. If any clear direction can be mapped, there has perhaps been a general trend away from simple, if rigorous, economic models towards more complex, but more subjective approaches: a move from confident optimism towards guarded pessimism and more introverted interpretations. Clearly, no single approach can be identified as the most incisive or the best for an understanding of relationships between migration and development. All have their strengths and weaknesses and all bring different perspectives to bear on the many facets of the complex interrelationships.

We have seen that the study of migration has been divided clearly into those studies focused on internal and those focused on international movements, with each coming from quite different traditions and using different types of data: the former coming from studies of individual countries in isolation in both the developed and the developing world and the latter coming from the historic movements out of Europe to the traditional centres of settlement in North and South America and Australasia. Yet the approaches used had much in common with theories based upon neoclassical principles, the new economics of risk minimization, segmented market theory and network analysis common to the study of both migrations. The increasing movements out of less developed countries to the more developed countries after the Second World War, particularly from the late 1960s, and the structural transformation of the global economy over the same period, urge us to consider internal and international movements within a unified framework. That framework needs to encompass both the spatial and the temporal dimensions of migration.

As migration is a process, present migration cannot be understood without reference to past migration. It appears almost inevitable that some form of transitional, evolutionary or historical approach is required if a reasoned analysis of that process is to be achieved. Whether the approaches need to be global, grand theories is not so clear-cut. In a previous book (Skeldon 1990), I attempted to show how variations in the mobility transition could occur but my emphasis was nevertheless on a single unilinear path of migration development. In this book, the global approach will be maintained, but the emphasis will be on regional patterns and migration transitions, in the plural, rather than on any single trajectory. The balance between local and global is necessary, but it would be unwise to reject generalizing transitional models outright. As Williamson (1988: 461) reminds us, in his wide-ranging review of the economic literature on migration and urbanization in Europe and the developing world, 'There is very little that is unique about Third World migration and city growth.' The existence of massive circulation during the process of urbanization, the development of so-called informal sectors, the fairly rapid incorporation of migrants into urban labour markets and the general improvement of their status all appear common to the experience of more developed and less developed countries. Given this situation, a distinction between the two may not be valid at all. However, although these parallels exist for some countries and regions, there are others where the experiences are quite different. In the next chapter, I will develop a framework based upon regional differences that will be used to view the apparent variety of pattern within a general system. These regions will be used for an analysis of migration across the world in subsequent chapters.

Chapter 2

Systems and boundaries

The idea of migration as a system incorporating a network of origins and destinations that change through time was raised in the previous chapter. Originally conceptualized for the analysis of internal migration, the system was assumed to be bounded by the nation state. It is also possible to examine subsystems of movement within any single country, and particularly in a large country, linking some regions more intensively together than others. When we come to international movements 'consideration of an all-encompassing global system would not be very enlightening' and separating one system from another becomes an important issue (Zlotnik 1992: 19). While this point is generally well taken, no system that can be identified at the international level will be entirely closed and there will always be 'spillage' into other systems. There may be some benefit in considering the world as a whole, at least to see not only how and why the major systems have evolved but also how they themselves interact, no matter how tenuously. Nevertheless, identification of system boundaries becomes a major methodological issue. Zlotnik (1992) would delimit the system on the basis of the strength of the linkages as measured by the number of migrants from country A in country B. A threshold therefore needs to be established to separate 'strong' linkages, included within the system, from 'weak' linkages, excluded from the system.

Boundaries plague just about every aspect of the social sciences. Classification into neat, mutually exclusive, Linnean categories is not usually possible, and the terms we use are often extremely imprecise. In this global examination of migration and development, it is not simply the delimitation of migration systems that is problematic. If there is a 'migrant', is there therefore a 'non-migrant'? There are also several different types of migrants, so how have these to be distinguished from each other? Given the importance of the transition to more urban societies and of rural-to-urban migration in that transition, the separation of populations into urban and rural sectors becomes an issue. Finally, there are the measures of development. The simple separation of developed from developing countries, so useful in general discussion, was questioned in Chapter 1. So many of our categories are in the form of binary separation: rural/urban; migrant/non-migrant; developed/developing; inside system/outside system. Yet are these simple categories really operational? If a more complex categorization is required, on what basis should it be made?

This chapter will attempt to review the criteria identifying the systems and boundaries used later in this book. This should give the reader an idea of just what is being discussed in each category and many of the usually implicit assumptions that underlie much of our analysis. I will focus on four general areas: international migration systems, regionalization, the separation of rural and urban, and migrant typologies. The first two categories might be thought of as essentially similar as migration systems are regions of a type; they are spatial units of more intense interaction of population. Yet, they may incorporate several different types of region as defined in a more conventional sense: that is, areas with some common identifying characteristics, whether cultural, economic or based on some other criterion or a combination of these. The discussion will commence at the macrolevel and move down to smaller spatial units before considering the variety of migrants and the channels through which they are moving.

The changing global migration system

The global international migration system has changed dramatically over the last 150 years. The second half of the nineteenth and the early twentieth centuries were characterized by great mobility dominated by movements out of Europe to North and South America and eastwards across Russia. In addition, there were significant movements out of China, also initially to the Americas, and to Australia, and later to Southeast Asia. The transatlantic movements covered the age of the 'Great Migration' and, starting as early as 1820, were really over by 1914 and had involved well over 50 million people from Europe (Baines 1991). This was a period when the world was dominated economically by Britain; and movements within the Empire, and within the colonial networks of other European powers, were important. These migrations and their relationships to development are outlined in Chapter 3.

The period between 1914 and the late 1950s covered the final stages of the European migrations and was marked by distinct fluctuations, when the movement of settlers was curtailed during the times of political and economic crises of the two world wars and the depression of the 1930s. The Second World War, and its immediate aftermath in Europe and elsewhere as countries broke colonial ties, saw vast movements of people. For example, it is estimated that 40.5 million people had been displaced by the war in

Europe by 1945 (Holborn 1968) and between 15 million and 17 million were involved in transfers from one country to the other upon the creation of independent India and Pakistan from British India in 1947. The millions who were displaced within the vast confines of China during the civil wars and the Sino-Japanese war can only be imagined, although the numbers leaving China upon the establishment of the People's Republic of China in 1949 were small compared to the total population – some 1 million to 2 million to Hong Kong and a similar number to Taiwan. The period between 1914 and 1945 has been called, with total justification, the 'age of global catastrophe' (Hobsbawm 1994).

After the disruptions of this period, migration began again from Europe to North America, but not on a scale or of a duration comparable to the period of the Great Migration. The 1960s can be seen as a watershed after which Europe faded as the major source of migration and itself became a destination of migrants from other areas. The international migration system became much more complex, involving many other, and often new, countries. Thus, the most developed countries of the world are facing the migration of large numbers of settlers from poorer, less developed regions who are bringing different cultural values and expectations. This recent migration is not the first time that migrants from vastly different cultural backgrounds and poorer areas have gone to the more developed countries. Hundreds of thousands of Chinese, in particular, went to North America and Australasia from the 1850s, but they saw themselves and were seen by others to be sojourners who would eventually go back home. Many did not, and their increasing numbers caused the destination countries to erect, in the words of Charles Price (1974), 'Great White Walls' and exclude not only Chinese but also most Asian and other non-whites from going to the United States, Canada, Australia and New Zealand. The infamous Exclusion Acts of the United States were implemented first in 1882, and extended until 1943, but even after their dissolution there was no real impact on migration until the major policy shifts of the 1960s, which introduced the new international migration systems.

Dividing the world, for the moment, into a simple twofold division of developed and developing countries, or North and South, as has become fashionable (see the Introduction), we can see how the pattern of migration has changed since the 1960s. The percentage of immigrants admitted to Australia, Canada and the United States which came from developing countries increased from 7.8, 12.3 and 41.1 per cent respectively in the early 1960s to 53.7, 70.8 and 87.9 per cent respectively by the late 1980s (Zlotnik 1991: 20). The absolute number of immigrants from developing countries increased some sixfold to Australia, sevenfold to Canada and fivefold to the United States over the same period. The case of migration to Europe was somewhat different, with only some countries, notably Germany and France, dominated by migration from the developing world, while others such as Belgium, the Netherlands, Sweden and the United Kingdom still took the majority of their immigrants from other developed countries. In almost all European countries (the United Kingdom was a notable exception), however, the relative proportion of immigrants from developing countries was increasing, and in some cases markedly. For example, in the case of Sweden, less than 5 per cent of its immigrant intake in the early 1960s had been from the developing world, and this proportion had increased to over 65 per cent by the late 1980s (Zlotnik 1991: 22).

These simple proportions and numbers, which give the impression of a generally increasing migration out of the less developed world to the more developed world, obscure significant regional differences. Just over half of all the 904,292 immigrants admitted to the United States in 1993 (including IRCA legalization)[1] came from just seven areas: Mexico, China, the Philippines, Viet Nam, the republics of the former Soviet Union, the Dominican Republic and India (INS 1994). Of the settler arrivals to Australia, some 49 per cent in 1993–94 came from six countries: New Zealand, the United Kingdom, ex-Yugoslavia, the Philippines, Hong Kong and Viet Nam. The origins of immigrants to Canada are a little more diverse, with Hong Kong accounting for 15.4 per cent of the total in 1992 and, together with the Philippines (5.2 per cent), Sri Lanka and India (5 per cent each), Poland (4.7 per cent) and China (4.1 per cent), constituting almost two fifths of the total intake. In the case of both France and Germany, over half of the stock of the foreign population in 1990 came from just three countries; Portugal, Algeria and Morocco in the case of France; and Turkey, ex-Yugoslavia and Italy in the case of Germany.

Such gross figures are equally deceptive as, clearly, the demographic giants of China and India can affect the total picture of movement. On a per capita basis, there is far more movement out of tiny Singapore to Canada, Australia or the United States, for example, than there is out of either China or India, which both figure prominently in the gross flows. The size of countries, as well as their relative location, influences the pattern of movement. Propinquity, as implied in Ravenstein's first 'law' of migration, is still an important factor, with Mexico being a major source of movement to the United States, and the southern European and North African countries being strongly represented in the movement to France. Yet, size and location are obviously only part of

---

[1]For a brief discussion of IRCA, see page 78.

any explanation. Britain, with the exception of migration from Ireland, has in the past attracted few people from areas close by compared with sources in South Asia and the Caribbean. Among the flows to the United States from Asia, the Philippines figures prominently, yet the much larger Indonesia does not; Thailand does, but neighbouring Malaysia does not.

Sassen (1988), in a wide-ranging review of international migration, associated the flows with the history of involvement, political and economic, of potential destination countries in areas that later became major sources of migrants. The flows to the United States from the Philippines, South Korea, Viet Nam, Taiwan, El Salvador and the Dominican Republic can be seen in this light. The legacy of colonial possession must obviously be a part of an explanation of movement to the United Kingdom or to France. But there has been little American involvement in India and China, which are major sources of movement to the United States. And what of major destinations such as Canada and Australia, which have pursued less adventurous foreign policy initiatives? The migrations into Europe in the 1990s, unlike the earlier movements, demonstrate that they can develop 'without geographical or colonial ties between the sending and the receiving countries' (Golini, Bonifazi and Righi 1993: 70). However, as Sassen makes clear, it is not simply political involvement that is of concern but the influence of the movement of capital through networks of transnational corporations, essentially the penetration of capitalism. This brings economic development and the evolution of a global system back into the foreground. Political involvement and the movement of capital are clearly interrelated, and the nexus of these influences forms the driving force of so many internal and international movements today.

The shift from a transatlantic to a transpacific system of migration and the establishment of south-to-north movements in the Americas and towards Europe should not give the impression that these are the only major international migration systems. From the 1970s, large flows of skilled and unskilled labour have been directed to the oil-rich countries of the Persian Gulf. Initially from neighbouring countries in the Middle East, the migration fields rapidly extended to include countries in South and Southeast Asia. These movements will be examined in Chapters 6 and 7. More recent changes show a slowing in these movements as increasing numbers of migrants move to the boom economies of East Asia. Thus, the boundaries of the systems are constantly changing as the migration fields are reoriented towards new areas of economic dynamism. The diffusion of Japanese capital is important in establishing linkages that give rise to increasing intra-Asian migration.

The state in the international system

In the analysis of the changing international migration systems we are plunged into one of the fundamental difficulties plaguing any analysis of global systems or of phenomena at the global level: the tension between the states on the one hand and the other actors in the global system on the other. The state remains one of the most powerful players in this system, even if its power and effectiveness vary from place to place. In the area of migration, state institutions are critical in conditioning many forms of internal population mobility, in the location of schools and education policies and in the size and recruitment policies of the military, quite apart from national development plans, which may affect industrial location, land zonation, and so on. State policies towards those who enter and leave are critical in determining types and patterns of international migration and have been considered the key variable which identifies internal and international migrations as two different mobility types (Zolberg 1978). State policies about who can and who cannot be members of that state are at the root of so much forced migration.

Finally, and perhaps most critical from an analytical point of view, the data upon which we depend are normally available for nation states and it is usually exceedingly difficult either to aggregate or to disaggregate the figures to create new spatial units. Aggregation immediately raises the issue of comparability of data and definitions from one country to another, and disaggregation can rarely be made for units ideal for the analyst. Exceptions do have to be made for the wealthier countries, where detailed geocoding systems may exist and where attempts at standardization are being made but, for most of the world's population, aggregation and disaggregation of data to units larger or smaller than the state remain a very real problem. The increasing trend towards regional cooperation underlines the need for aggregation of data. The European Union is the most obvious example, but there are growth triangles or other transnational zones or regions of less determinate geometry in several parts of the world (see Chapter 5). Conversely, states do not usually form homogeneous units and there are considerable differences in level of development within countries – if it were not so, then internal migration would not be as important as it is – and these differences are likely to be more significant the larger the country. The disaggregation of data into development regions, how ever defined, becomes very important in these cases. Thus, the problem of regionalization exists on several scales, transnational as well as national. Very generally, regions are created by two very different sets of forces. The first is internal to the region, including essentially geographical and cultural factors which provide some kind of identity to the region; the second set of forces is external to the region and comprises those which provide it with a location relative to prevailing centres of power. The following discussion focuses on the measures used to create a series of development regions that provide the framework for the analysis of migration and development that is used in the subsequent chapters.

Regionalization

The task of dividing the world into coherent development regions immediately raises the issue of the criteria that are to be used for such a division. There are as many ways of dividing the world into regions as there are favoured criteria and ideological background: economic, political and cultural variables can all be used. If one is to believe that the direction of future world development will be determined by the 'clash of civilizations' (Huntingdon 1993), then culture worlds such as western and eastern Christian, Islamic and Sinitic regions need to be identified. As will be seen so often in subsequent chapters, cultural factors are important in explanations of migration, but regions based upon culture are likely to be too broad and heterogeneous for all but the most general of discussions of development. The legacy of differences between Catholic and Protestant areas in a western Christian region is surely not entirely irrelevant, and the whole debate about Confucianism and Asian values and development is driven more by national sentiment than by objective analyses. A map of world political systems could be constructed to identify liberal-democratic regimes, socialist systems, theocratic systems or military dictatorships or, perhaps more usefully, into 'hard' (strong) and 'soft' (weak) states. The political dimension has often been sidelined in analyses of development compared with the economic dimension. Weiss and Hobson (1995) clearly show the critical role that strong states have played in the historical experience of Europe, and continue to play in the world today, even if the factors making a state strong today differ from those in the past. Constantly shifting political systems, however, make the mapping of such regimes somewhat problematic, although the structure of the state must be incorporated into any attempt to create development regions.

Economic criteria incorporating the gradual change of variables over time are still arguably the 'best' measures of development and, despite all the drawbacks, GNP per capita is still perhaps the most commonly accepted indicator. GNP is a reasonably objective measure, with a common methodology of calculation across countries. GNP per capita is now enshrined in the development literature and, through annual updating, is a foundation for the ranking of countries in the *Development reports* of the World Bank. There are, of course, weaknesses with the approach, not the least of which is the conversion of all countries' income measures into United States dollars in a way that can give an idea of the 'real' value of that income within each county. A way round this problem is to give an idea of the purchasing power that each currency commands within each economy through a purchasing power parity estimate of GNP per capita. The standard GNP per capita measure, however, remains the basis for World Bank country classification, and several variations have been proposed, mainly to incorporate more socially oriented development variables. For example, the physical quality of life indicator and the more recent variant, the human development index, of the United Nations are attempts to emphasize the importance of non-income variables in any assessment of development. The human development index is a composite index of GNP per capita, adjusted for purchasing power parity, life expectancy at birth and a measure of educational attainment. The 1995 version of the *Human development report* published by the United Nations (1995a) included a gender-related development index to complement the human development index and highlight gender inequalities across nations. Some economists argue that these variables are not independent and that we are still best served by the single GNP per capita measure despite its weaknesses (Kelley 1991).

A map of countries by GNP per capita is given in Fig. 2.1, divided according to World Bank criteria into low-income economies, lower-middle-income economies, upper-middle-income economies and high-income economies. All but the last category fall under the World Bank's definitions of 'developing economies'. Such a simple division immediately raises the tensions outlined above between the state, or the area over which governments have control and on which universal basis all population and development statistics are released, and the underlying 'real' spatial pattern of development. There are major differences in level and type of development within countries that are obscured by the even shading of the map and which are often ignored when the statistics appear as listings in tables. The larger the country, the greater these differences are likely to be. There are functional divisions of economy over space which have extended and become more complex with the trends towards a global economy. Rapid transnational communications have allowed companies to take advantage of factor prices at different locations around the world. Labour-intensive parts of a production process can be moved from areas where rent and labour costs have become high to areas where they

Fig. 2.1 Groups of economies as identified by the World Bank

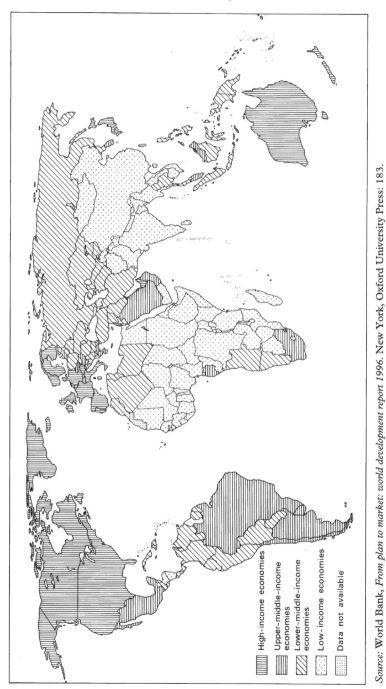

High-income economies

Upper-middle-income economies

Lower-middle-income economies

Low-income economies

Data not available

Source: World Bank, *From plan to market: world development report 1996.* New York, Oxford University Press: 183.

are low, while still maintaining the organizational and capital-intensive parts of the process in the original location. An inter-national division of labour has emerged wherein capital-intensive and service activities are concentrated in developed core locations while labour-intensive, mostly manufacturing, activities are moved to offshore sites where labour costs are low. The division is not a simple one between more developed countries on the one hand and less developed countries on the other; specialist zones are emerging in particular parts of the world that are based around urban centres, and these are coalescing into development zones or corridors. These zones typify the more developed core countries as much as they do the less developed periphery countries, although their form and function may differ from the one to the other.

It is in the superimposition of these zones onto the national patterns identified from the World Bank data that difficulties arise, often because of a lack of data, but also because the zones themselves have no hard-and-fast boundaries. An attempt has been made in Fig. 2.2 to map a series of 'development tiers' that tries to capture the idea of a transition in types of economies from those with essentially primary agricultural or extractive activities, through those based upon labour-intensive industries, through those centred around capital-intensive, wholesale and retailing activities, to those based on knowledge-based industries. At the global level, these development tiers can be but a highly generalized description of the kinds of activities found in each region and they cover a multitude of local variations. They do, however, shift attention partially – and it can only be partially – away from the state as the only spatial unit of analysis. In subsequent chapters, these development tiers will be used as a framework to examine the types and patterns of population movement both within and between the tiers. I argue that it is through the use of such a framework that we can better capture the interrelationships between development and migration.

The state, nevertheless, must remain central to our analysis. One could argue that, as development implies planned change, it must by definition be state-centred: the state is the only unit that can directly implement development policy. This is not to say that all development is brought about by the state but that, despite the rise of multinational corporations and other transnational regional and global institutions, we are still very far from witnessing the end of the nation state (Ohmae 1995). States exercise control over people and territory, and this is most effectively realized through the penetration of institutions or infrastructure. Although states can be held together by authoritarian governments exercising despotic control, economic development is fostered only when there is infrastructural penetration (Weiss and Hobson 1995,

Fig. 2.2 A schematic representation of a system of migration and development tiers

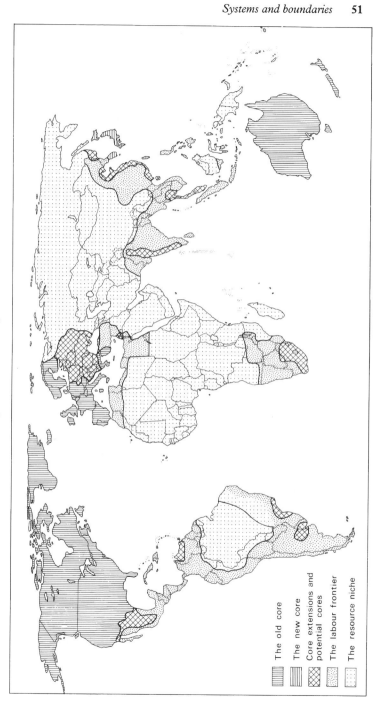

The old core

The new core

Core extensions and potential cores

The labour frontier

The resource niche

following Michael Mann). The institutions to effect this penetration are various and include economic (fiscal), military, construction of communications, of roads, railways and so on, as well as the
means of political participation. The state is articulated by circuits
of human mobility ranging from the more coercive such as con-
scription, through labour recruitment schemes to more apparently
voluntary movements to centres of education. By bringing peoples
from different areas together, the mobility fosters common languages for communication, and the military and education systems
mould common traditions and values. The centralizing requirement of these institutions, which locates the principal military and
educational training centres close to centres of government, often
conflicts with policies of decentralization whose stated aims are to
slow migration. Thus, there can be tensions between overt policies
to control population movements and the unstated, and often
unrecognized, need to consolidate the nation state.

The structure and nature of the state are changing, too, over
time and space. The five development tiers in Fig. 2.2 also try to
capture differences in state structure as well as economic development. The first two tiers include those states which have total
infrastructural penetration and thus their boundaries correspond
with state boundaries. Parts of these states, too, may be beyond
effective control, particularly in certain neighbourhoods of larger
cities, but their geography is too complex to include in a highly
generalized global map. Class and ethnic enclaves will also generate their peculiar patterns of population mobility. The effectiveness of the state, as defined not simply by political control but by
the efficacy of infrastructural penetration, declines progressively in
the other tiers. The thrust of the following chapters will be to
demonstrate that there is a relationship between the level of
economic development, the degree of state formation and the
patterns of population mobility. Very generally, we can say that
where these are high, an integrated migration system exists consisting of global and local movements, whereas where these are low the
migration systems are not integrated and are mainly local.

There are general, if variable, trends towards decentralization
of populations and for immigration in the first two development
tiers, with the pressures being more intense in the first tier; a
general trend towards the centralization of populations and both
immigration and emigration in the third tier; in the fourth tier,
there is also a tendency towards centralization, but the dominant
characteristic is emigration; and, finally, we find a very variable
pattern of migration in the fifth tier, including emigration and
immigration, centralization and decentralization. Thus, the diagram attempts to integrate, in a very general way, economic,
political and migration structures. The basic demographic and

economic variables for countries and areas allocated to the development tiers are presented in Annexe Table 1, which clearly reveals the practical problem of attempting to rearrange national-level data in any system not based upon states.

The first two tiers identify areas that can be considered developed: the 'old core' areas of western Europe, North America and Australasia and the 'new core' areas in East Asia. There are obviously major subregional development areas within these broad tiers, such as the 'blue banana' economic heartland area in Europe (Rimmer 1994) but, at a global level of analysis, not all subregional variations can be considered important. These subregional development corridors spread from global cities, which are the linchpins at the pinnacle of the global system of migration and development. The global cities coordinate the functional division of labour and of space down through an urban hierarchy. Global financial centres such as New York, London and Tokyo are at the apex – all global cities will be found in the old core and new core tiers – and these are linked to the other tiers in a system of nested urban hierarchies. The evolution of this system of global cities has been traced by King (1990) and by Chase-Dunn (1985). They have shown how these urban systems have expanded from the core into the periphery over the last 150 years. The third development tier, which I have called the 'actively expanding core' is widely scattered and subdivided into three broad classes of actual and potential expanding cores and restructuring cores: in coastal China and central parts of Southeast Asia; in parts of western India; in southern Africa; around Israel in West Asia; in four parts of Latin America; and in western Russia and eastern Europe. Away from these areas I have designated a fourth tier of a 'labour frontier' and, in the most isolated areas and in small economies, a fifth tier representing a 'resource niche'.

It is important to bear in mind that the boundaries between the development tiers are largely notional; they are more indicative of general areas of different types and levels of development than precise lines on the ground separating clearly distinct regions. The boundaries are also constantly shifting over time and space, although these shifts should not be envisaged as any regular or sequential trend of diffusion in any particular direction such as a series of frontiers passing in waves of increasing intensity of human use of the landscape as envisaged by Turner (1894) in his frontier hypotheses of American development. As Storper and Walker (1989: 208), who involve the spirit of Turner in their work, remind us, the expansion of capitalism is inconstant, vaulting across space to favoured regions, leaving a complex variegated pattern that renders simple regionalization problematic.

The delimitation of these development tiers should not imply that these regions are necessarily homogeneous. They are not and, in the chapters that follow, important differences will be drawn within the tiers themselves. Given the problems with the delimitation of the development tiers, it might be thought that the simple distribution of states might be satisfactory as an approach. The advantage of the proposed approach is, first, that it moves away from the simple binary distinction between developed and developing countries, convenient though this may be. All countries are developing in one form or another, and it is important to envisage these within a single framework. Second, this system of regionalization, unlike the simple distribution of states, brings to the fore the dynamic relationships in the global economy and society, between economies, politics and mobility, and there is a functional relationship between regions that changes over time. Finally, although global linkages and relationships are being emphasized, the approach, as we will see, also draws attention to the tenuous and fragile nature of globalization. Some parts of the world are much more closely integrated into the global system than others and some may even be 'disarticulating' from that system. There is nothing inevitable about globalization and the boundaries between the tiers can move either way.

Attention has already been drawn to subregional divisions that exist within the development tiers. These can be identified as areas of greater or lesser development or – of critical importance in the analysis of migration – by sector, urban and rural.

Rural and urban boundaries

The growth of cities and the transfer of population from rural to urban areas have been among the most studied of topics. Yet, precisely what we mean by the city, or the creation of a definite boundary which will separate a rural sector from an urban sector, remains elusive. In part, this difficulty stems from the fluidity of the real situation; there is continual interaction between urban and rural and any attempt to delimit them into separate sectors will be artificial. However, as in the case of the development regions discussed above, some separation is required in order to allow an understanding of the processes which are operating to transform the society.

A major and well-known problem relating to any comparative analysis of intersectoral migration is that the definitions of urban are not consistent across countries. Definitions are generally based on one of three criteria, or sometimes a combination of these: administrative, population size and/or density, and the proportion of the labour force in non-agricultural activities. Those based upon administrative criteria depend upon legal definitions of towns or municipalities and are usually the areas controlled by some form of urban local government. These urban centres may include large areas which are rural or, given the inertia of much of government, may have been fixed for some considerable time, while more recent urban growth has occurred outside the municipal area. Thus, urban, as defined by administrative criteria, can overstate (for example, China) or understate (for example, Thailand) 'real' urban development.

Size and density functions depend upon particular thresholds being set to define what the size and population density of contiguously built-up settlements should be to be defined as 'urban', and these vary from country to country. Rarely, however, do size and density thresholds capture the 'essence' of urban and often an 'urban characteristic' requirement is also added. This could be a municipal government or some infrastructural characteristic such as paved roads, piped drinking water supply, electricity or reticulated sewerage systems. Even in the case of quite large settlements, significant proportions of the population may be engaged in agricultural activities, commuting to their fields every day as in the Mediterranean world. A threshold of a proportion of the population engaged in non-agricultural activities may therefore provide a more useful definition of the 'real' urban population.

Not only is it difficult to compare definitions of urban in one country with those in another, but the definitions within a single country can change over time. One of the most complicated cases is China, where estimates of the urban population have varied between 23 and 50 per cent and back to 25 per cent of the total population within a few years. There is no alternative to a painstaking examination of all the criteria used to define urban areas in each census, survey and official set of registration figures in order to bring some degree of comparability to the data. For exemplary attempts for China, see Chan (1994) and Chan and Xu (1985).

An even greater problem relates to changing urban boundaries. Even where the definitions of urban remain consistent over time, the boundaries need to be changed to reflect real urban growth. However, that growth often incorporates densely populated rural areas, and often also smaller urban centres, which previously lay outside the expanding town. To be sure, the incorporation of these areas does indeed represent the real growth of the settlement. Unfortunately, the details of the additions are rarely made prominent in census reports and thus the growth of the town over time is based upon the comparison of two quite different geographical areas. It may even be impossible to reaggregate present or previous census data to make the spatial units identical.

For one attempt to reaggregate census data for Sri Lanka, see ESCAP (1980).

The above points may appear to be somewhat laboured and esoteric, but they are quite critical for the analysis of migration. There are three components in the growth of any urban area: natural increase, or the balance between births and deaths; net migration, or the balance between inmigration and outmigration; and reclassification, or that population added (or subtracted) through the addition (or exclusion) of any area. The calculation of the components of urban growth between two time periods is normally on the basis of projecting the population of the urban areas forwards on the basis of births or deaths or on the basis of the overall national census survival ratios (see Zachariah 1977 for a discussion of the methodologies) to produce an expected urban population at the later time period. When that expected figure is compared with the population actually counted, any difference is assumed to be due to net migration. Thus, reclassification tends to be included in the net migration component, which can artificially inflate the role of migration in urban growth. Where disaggregation of the data has been possible, reclassification has been shown to have been a very significant proportion of urban growth in certain countries at certain times (ESCAP 1980). Because of the difficulty in computing reclassification, many analysts choose to ignore it (Ledent 1982, Rogers 1982, Berry 1993), leading to somewhat spurious interpretations or uncomfortable questions about hidden factors in urban growth. Unfortunately, in a global examination of migration, one has to depend to a large extent on the available macrolevel data, but students must be warned that the country data really need to be carefully examined in their own right before detailed analyses are possible of the role of migration.

While the expansion of urban boundaries can cause analytical problems, so too can the converse: a lack of change in urban boundaries. Hence, the same area may be compared over successive time periods but real urban growth beyond the boundaries is excluded, thus deflating the urbanization level, urban growth and the role of migration. Also, if the boundaries of large cities are not extended outwards to incorporate fast-growing settlements on their periphery, a superficial analysis of the growth of urban places may show that medium-sized or small towns are growing fastest. In reality, these settlements are but suburbs of the metropolitan centres. Thus, what might appear to be decentralized urban growth in smaller towns is, in reality, continued centralized growth in the largest city. The spatial location of urban centres in the hierarchy always needs to be appreciated in any examination of the overall pattern of urban growth.

There are several other problems in delimiting urban and rural populations, the most significant of which relate to the method of registering populations at their places of residence. Censuses and surveys generally use either a *de facto* system, where persons are recorded where they actually are at the time of the census or survey, or a *de jure* system, where persons are recorded where they usually live at the time of the census or survey. As I have made clear elsewhere (Skeldon 1987a, 1990), a *de facto* method is to be preferred as this will capture a higher proportion of total mobility, including short-term movers, than the *de jure* system. *De jure* censuses tend to underenumerate the population of large urban agglomerations (as they have in Thailand, the Philippines and Indonesia, for example), because they enumerate many urban migrants in their rural places of origin rather than in the cities. Many migrants may be in the city only for a short time, but they are constantly being replaced by other short-term migrants, and this sector of high turnover is a permanent feature of the urban population which also needs to be supplied with basic services.

There are thus very real practical difficulties in trying to impose a boundary somewhere along the rural-to-urban continuum. While there are clearly urban and rural places that we need to separate for analytical purposes, the very existence of rural-to-urban population movements renders the separation of urban and rural systems problematic. As with the previous case of regionalization, the boundaries are largely notional and of analytic convenience, rather than representing any hard-and-fast separation of systems.

Typologies of migrants

The separation of migrants into separate types or groups is also an exercise in the creation of systems and the imposition of boundaries. All people move during their lifetime, but there is a world of difference between those who have moved only within the local area and those who have spent years overseas. Three dimensions are usually employed in categorizing migrants and migrations: space, time and motivation (or purpose). Once again, a series of dyads or binary divisions is often used: long-term and short-term; permanent and temporary; long-distance and short-distance; forced and voluntary; legal and illegal. Yet again, thresholds need to be imposed, particularly along the time dimension. More complex divisions can be introduced such as the difference between daily, periodic, seasonal and long-term circulation, and irregular and permanent migration proposed for the classification of African population movement, for example (Gould and Pro-

thero 1975). Within these space–time divisions, the authors pro-
pose motivational categories revolving around a basic economic/
non-economic dyad, with further subdivisions within each of the
binary groups.

While patterns of daily, weekly and seasonal movements are
generally well established among virtually all groups, difficulties
emerge in trying to distinguish between circulation and migration.
A movement that started out as a short-term visit to a city may
eventually turn out to be a more-or-less permanent movement.
The idea of a settler as a permanent migrant is often contradicted
by the evidence of high rates of return or onward migration, even
after relatively short periods of time as a settler. As will be seen in
Chapters 3 and 4, the distinction between settlers as permanent
migrants and sojourners as more temporary or circular migrants is
often blurred. It is even too simple to suggest that a movement only
becomes truly permanent once a person is dead, as the migration of
the dead is a not insignificant, although largely unacknowledged,
phenomenon. Burial societies among migrants are widespread in
cities in many, very different cultures and these ensure the repatriation of bodies to home areas. The transfer of the bones of the
ancestors may indicate a permanent shift overseas from China; and
the reburial of those killed in recent fighting in the republics of the
former Yugoslavia reinforces the ethnic integrity of particular
areas.

In practice, time limits are imposed on the data to separate
movements of longer and shorter duration. Distance is even more
problematic and is normally dictated by the size of spatial units
used to measure migration, which are the given administrative
units of a country. Movements within districts, for example, can
be assumed to be local, while those from one state to another can
be taken to be more long-distance, though this need not necessarily be the case. A movement just across a state boundary would
obviously be a 'local' move, for example. These issues are well
illustrated in existing texts on population or migration (Jones 1990,
Parnwell 1993), and there are almost as many typologies as there
are varying research objectives among those analysing human
mobility. For an early classic example, see Petersen (1975), and
several examples will be noted in the chapters that follow. The
categorization of international movements, probably because most
are assumed to be of longer distance and longer duration, tends to
focus more on the motivational or functional dimension. The
available immigration data are also often categorized by type of exit
visa or immigrant class of entry, whereas census data for the study
of internal migration can rarely identify the type of person moving.
In international migration it is easier to separate those going to
settle from those going on a temporary labour contract and those

on student visas. Hence, settler, labour, student and asylum-seeker migration systems have been identified (Skeldon 1992a), and there are many variants of this approach (Appleyard 1989). Again, the boundaries between them are porous: students can become settlers or asylum-seekers, or settlers can return home, for example.

The dyadic division between forced and voluntary migrations is common in many approaches. Slavery is a clear-cut example of a forced migration. In addition, the threat of persecution if a person returns to his or her homeland would appear to be a fairly clear way to identify a refugee. But, here again, the boundary is not necessarily so easy to draw. Refugees can become 'economic migrants' or 'illegal migrants' at the stroke of a pen as host areas determine whether people leaving a country are really fleeing persecution or are simply trying to take advantage of an available 'channel' leading to opportunities in other parts of the world. Refugees are thus redefined by the international community on the basis of many factors, not the least of which is political expediency as refugee fatigue increases and domestic populations agitate against the easy entry of many unskilled people who may have to be supported by their taxes. More restrictive immigration policies may be the result (see Richmond 1994).

Conversely, many of the 'voluntary' movements may not be entirely the result of free choice. Draftees into the military are an obvious example, but the importance of company transfers or the pressures brought by labour recruiters blur any clear distinction between what is truly free choice and more controlled movements. Thus, free movements may not be so free and forced movements may not be so forced, although we must always accept that there are indeed forced migrations based upon violence. A way round the problem of single binary divisions is to focus on the channels through which people move, identifying the agencies and agents involved in facilitating the transfer of people from one area to another (Findlay 1990). This is a variant of a migration systems approach at a more microlevel, although here again the delimitation of boundaries clearly marking the channels can be problematic. What is clear is that, as the global migration system has evolved, the channels of human movement have become more numerous and more complicated.

A final and increasingly important channel that can be identified is that of illegal or undocumented migration, as opposed to all the previous types of movers, who go through official channels. Indeed, here the boundary appears to be tight, although unfortunately we can rarely find out how many people fall into this category as, by definition, they are beyond the official data-gathering net.

Discussion

This chapter has drawn attention to the way in which the subject areas of migration and development can be disaggregated into categories for analysis. I have focused on the issues of regionalization, sectorization and the division of migrants into systems and types. In the following chapters of this book, I use the system of regions, and of sectors within regions, to examine the migration systems and types of migrants in each part of the world. Nevertheless, as this chapter has clearly shown, the boundaries that we use to separate spatial and sectoral units and to define the classes of migrants themselves are porous and are merely useful heuristic devices to help us to make sense of a very complex reality. Even if we could fix the boundary of a development region or an urban sector, it, like the concept of optimum population, would have to shift continuously, given the constantly changing reality. As was clear from Chapter 1, as our ideas about migration and development have changed over time, so too have the interpretations of our classification of migrant types such as settler or refugee.

In the framework outlined above, it is the migration system that links the regions and sectors together. Clearly, there are other flows that bind the regions and represent systems of interaction. For example, foreign investment or trade flows would be two of the most important economic factors linking the global economy. These are not independent of population movements, but their relationships with them are ambivalent. Whether trade or investment, by stimulating production in particular areas, acts to slow outmovement by improving conditions in origin areas, or accelerates migration by drawing people in to particular development nodes and increasing aspirations, thus encouraging more people to seek opportunities overseas, remains unclear. The information in the following five chapters, each dealing with one of the development tiers, will, I hope, throw some light on these and other issues relating to migration and development.

Chapter 3

The old core

Irrespective of whether we use simple GNP per capita as an indicator or some broader index such as the human development index of the United Nations, groupings of countries emerge in which western Europe, North America, Australasia and Japan can be identified as the areas of highest development. Apart from their general affluence, these areas also have three other common characteristics that are important for our discussions of migration and development, even if there are variations between them. The countries are all governed by some form of parliamentary democratic system, all are societies of low fertility and mortality and all are among the most highly urbanized parts of the world. While there are other common characteristics ranging from the low proportion of GNP that is derived from agriculture to the importance of the rule of law, the differences between these areas are striking.

The United States, Canada, Australia and New Zealand, as nations, were established as offshoots of Europe. They are young settler societies created through migration from Europe, mainly from the seventeenth century, in contrast to the long history of settlement and development of European societies, which dates back several thousand years. Of course, the history of settlement in North America and Australasia can be traced back many thousands of years too, but the displacement and marginalization of 'first' Americans or Australians raise large some of the inherent contradictions in any discussion of migration and development. The emigration out of Europe that was fundamental to the later development of the settler societies was the antithesis of development for the technologically simpler peoples with whom they came in contact, an issue to which I will return in the Conclusion.

Europe and the settler societies, as the developed old core expanding into the periphery of the rest of the world, give a very Eurocentric bias to world systems theory. Large parts of the world were initially, however, only superficially affected by Europe. China, the most developed society and economy in the world until well into the eighteenth century, certainly did not consider itself to be on the periphery of anything: it was the core, the Middle Kingdom. However, the overseas expansion of China, while it was establishing trading-posts and small colonies, never gave rise to mass movements of people equivalent to those from Europe. The state has actively discouraged emigration throughout much of its history. It was an island nation on the periphery of that Chinese core, however, which embarked upon a revolution that within a century was to bring it among the ranks of the most developed countries of the world. The migration experience of Japan, while exhibiting some parallels with European countries, has nevertheless been very different. The comparative relationship between migration and development in Japan, the first developed country outside the European, or European-generated core, is particularly germane. Japan and the other areas of recent prosperity in East Asia will be considered separately in Chapter 4 on the new core tier. In the present chapter, I will trace the changing patterns of population migration within the old core areas up to the present day, showing how current migration is evolving within a global system of migration.

The historical background to migration in Europe

The contrast between a settled Europe and a settler North America or Australasia is both true and deceptive. One of the great myths about migration is that pre-industrial or pre-modern societies were immobile. The idea that peasants are as 'fixed as the soil from which they draw life' (Bowman 1916: 63) has proved remarkably resilient in the interpretation of traditional societies in both Europe and Asia and probably derives from the popular contrast that is drawn between the settled agriculturalist on the one hand and the nomad on the other. Yet, all the evidence points towards a very high degree of population movement in pre-industrial societies. I briefly reviewed the evidence up to the late 1980s in a previous book (Skeldon 1990: 27–46) but, since then, has come Moch's (1992) superb review of migration in Europe since 1650. Following fellow historian Charles Tilly, she identified four main types of migration 'according to the distance of the move and the definitiveness of the break with home' (Moch 1992: 16–17): (a) local migration, which encompassed movement within home market areas for marriage or work; (b) circular migration associated with seasonal employment beyond the immediate areas; (c) chain migration, which channelled people to particular destinations, usually urban centres; and (d) career migration, which was controlled by institutions, especially the Church and later state bureaucracies, rather than village or family networks. These four 'free' migrations dominated the European mobility system until 1914. Although they overlapped, Moch (1992: 18) observed a progressive shift from the dominance of local and circular types in the seventeenth century to the proliferation of chain and career migration in the nineteenth and twentieth centuries. Although this trend was observed without reference to a mobility transition, it would certainly fit a modified version of that concept. Yet, the evolutionary separation of the migration types cannot be pushed too far, as Moch well appreciates. Even by the middle of the nineteenth century, much of western Europe was still dominated by systems of circular migration both to cities and to agricultural areas for harvesting (Chatelain 1976, Weber 1977, Devine 1979). Levels of urbanization in mid-century were, with the exception of England, still relatively low (Table 1) and the great age of rural-to-urban migration encompassed the second half of the nineteenth century and the early twentieth century.

At the beginning of the nineteenth century, the level of urbanization, as defined by the proportion of the population in settlements larger than 5000, for Europe (excluding Russia) was about 12 per cent, roughly what it had been a century earlier. By

Table 1.   **Levels of urbanization in the principal developed countries, 1800–1980**

	1800	1850	1910	1950	1970	1980
Belgium	20	34	57	64	71	70
England	23	45	75	83	81	79
France	12	19	38	48	68	69
Germany	9	15	49	53	68	75
Italy	18	(23)	(40)	(56)	65	65
Netherlands	37	39	53	75	83	82
Portugal	16	(16)	16	25	29	34
Romania	7	(11)	16	28	47	56
Spain	18	(18)	(38)	(55)	70	73
Sweden	7	7	23	45	62	64
Switzerland	7	12	33	48	59	58
Yugoslavia	10	(10)	10	16	37	44
Europe	12	19	41	51	63	66
USSR	(6)	(7)	(14)	(34)	54	61
United States	5	14	42	57	66	65
Canada	6	8	32	46	56	58
Australia	–	(8)	(42)	59	79	80
Japan	(14)	(15)	18	38	72	78

Source: P. Bairoch (1988) *Cities and economic development: from the dawn of history to the present.* Chicago, University of Chicago Press, p. 221 Table 13.4.

Notes: 1.   The fact that these figures have been only slightly rounded off does not imply a correspondingly small margin of error. The figures in parentheses have a much wider margin of error than the other data.
2.   A criterion of 5,000 is used for urban population.
3.   Figures for 1800 are very approximate.

1850, this proportion had reached 18.9 per cent and by 1900 it was 37.9 per cent (Bairoch 1988: 216). These bald proportions obscure the fact that large numbers of those recorded as living in towns were essentially rural-based circulators and thus, as discussed in Chapter 2, any separation of urban and rural sectors is largely artificial (see also Langton and Hoppe 1990). The principal conclusion to be derived from the recent research on migration in Europe at this time is that the great complexity of human movement that has been observed in many parts of the developing world today had its counterpart in the shift from a rural to an urban society in Europe in the nineteenth century.

Transatlantic movements: the early pattern

There were two main periods of migration out of Europe to the Americas. The first was mainly from Britain and followed a period of sustained population growth that terminated in the economic and political crises of the mid-seventeenth century. In the space of only eleven years, from 1629 to 1640, some 80,000 people left England for North America, with another 20,000 going to Ireland, and equal numbers going to the Netherlands and the Rhineland, and to the West Indies (Fischer 1989: 16). The majority of those going to North America at that time went primarily to Massachusetts, and they came from the eastern counties of England. They were followed by three other distinct streams: from the southern counties towards Virginia from 1642 to 1675; from the northern Midlands towards Delaware from 1675 to 1715; and from highland Britain towards the backcountry of the American colonies from 1715 to 1775. As Fischer (1989) demonstrates, in a monumental study of migration, these four flows gave rise to four separate culture regions along the eastern seaboard of the United States that have persisted to this day.

This interpretation emphasizes roots and the importance of the background of the early migrants and their cultural baggage in the creation of American culture, rather than the role of the American environment or even the thousands of immigrants from more diverse backgrounds who were to follow many decades later. One might wonder how voluntary much of the early migration could have been at a time of economic and political crisis, as Fischer maintained it was, but the British migrants, unlike their French or Spanish counterparts, had the freedom to develop their communities independent of government control. There can be few other studies as exhaustive as Fischer's of the role of migration in the development and cultural crystallization of regions at destination areas.

A detailed analysis of every person officially known to have left Britain for America between December 1773 and March 1776 (that is, at the very end of the period covered by Fischer) shows such an extensive variety among the migrants that two completely different patterns of movement have been identified for that time (Bailyn 1987). One was from southern Britain, dominated by males who emigrated individually or in small groups and who had come directly or indirectly through the larger cities, mainly London. These included people who had been internal migrants moving to the cities without the intention of emigrating but who 'found in emigration a resolution of their problems which they had not originally intended to pursue' (Bailyn 1987: 117). The vast majority of these emigrants had some resources in terms of money or education, and they joined urban or rural labour markets in America. This is the 'metropolitan' pattern. The 'provincial' pattern, on the other hand, consisted much more of quite large family groups from farming backgrounds, although not from the poorest strata of rural society, who sought out the backcountry in America to pursue life on the frontier. There was a whole range between these two ideal types, but they seem to capture much of the essence of this early migration to North America.

These two examples of transatlantic migration demonstrate a feature that is common in the movements in many of the less developed parts of the world today: streams from distinct origins that reproduce themselves in clearly identifiable ways at the destination. They control access to particular activities; through associations, they create groups that facilitate adaptation to the destination and promote development, as well as further migration from the origin; and they establish distinct ethnic, or simply regional, cultures, in a simplified manner perhaps, at destinations that are different from the home area, while reflecting its basic characteristics. Finally, it is not the poorest who move but those who have some resources.

Transatlantic movements: the Great Migration

The second great phase of migration out of Europe coincided with its period of most rapid economic development and of urbanization. Between 1860 and 1914, over 52 million people left Europe, mainly for the Americas but also for Australasia (Moch 1992: 147). This was the great period of European overseas settlement. Yet, the image of the creation of stable settler societies again obscures the complexities of the migration. Many of those who went with the intention of settling returned, others went with the intention of returning at some future date, and still others were moving within seasonal patterns of labour migration to undertake specific tasks in the New World.

There are few data to separate the transatlantic migrants into definite classes of temporary or permanent migrants and, in fact, it may be impossible to do this. Baines (1991: 40) believed that it was 'possible that *most* emigrants expected ultimately to return.' He estimated that perhaps one quarter of all emigrants between 1815 and 1930 returned to Europe, with lower rates of return for northern than for western or southern European countries (see also Chan 1990). Nugent (1992: 35) cites rates of return from Argentina between 1857 and 1914 as 43.3 per cent, from Brazil between 1899 and 1912 as about 66 per cent and from the United States between 1908 and 1914 as 52.5 per cent. Many of the Europeans were thus sojourners – an important point to bear in mind when we compare this movement with that from Asia in the nineteenth century. Large numbers of Europeans too, like the Chinese, were young single males who had no intention of settling at the destination. They were temporary labour migrants. This particularly applies to the great surge out of Europe from the 1870s to 1914, which was heavily biased towards males. Previous movements had been roughly balanced in terms of sex ratio, and the small post-First World War migrations were female-dominant, though there was considerable variation by ethnic group (Nugent 1992: 155–6). The incidence of return for women was much lower than for men, reflecting a pattern that we will see in many other parts of the world in more recent times: the participation of women leads to a reduction of circulation.

These were the European 'settlers'. The Chinese, on the other hand, tended to be seen in diaspora always as having the intention of returning. Yet many stayed, and a crude estimate of return during the period 1848–82, when the Chinese were free to move backwards and forwards, has been given by Chan (1990: 38) as 47 per cent, clearly comparable with the European rates. There were indeed differences between European and Asian patterns of migration to settler societies but these were due mainly to regulations and restrictions implemented by the groups in power in the destinations rather than to any simple differences between the intention to go as settlers and the intention to return. As we saw in Chapter 2, Asians were virtually excluded from Australasia and the Americas from the 1880s, a situation that was not to be reversed for over eighty years. However, as we will see in subsequent chapters, many of these issues of return migration are as pertinent to today's migration as they were to the movements of a century or more ago.

Much of the transatlantic migration appears to have been

merely a spatial extension of European systems of circulation: a circulation allowed by improvements in transport. Thus, although pre-industrial Europe had been characterized by a high degree of population mobility, the economic and technological development from the mid-nineteenth century extended the types and the spatial range of that mobility. Although systems of circulation persisted in both the internal and the international European migration systems, the general trend within Europe at that time was towards the concentration of population in urban areas. In the movement to the Americas, there was both concentration and deconcentration.

One interpretation of American history is to see the nation as the product of the frontier, of the gradual westward movement of its population conquering a new wilderness environment in what is essentially a complex system of multiple moves (Turner 1894). The reality was different. The occupation of rural lands was only one part of the story, and many immigrants settled in the urban areas along the eastern seaboard and into the mid-west in what can be seen as an extension of European urban-to-urban or rural-to-urban movements. The urbanization level of the United States increased from 14 to 42 per cent between 1850 and 1910. Very generally, movements before the 1890s were biased towards the search for agricultural land, while those after the 1890s contained more urban labourers, with greater tendencies to return to Europe. That distinction represents only the most general of trends as migration for wage labour had begun as early as the seventeenth century, as Bailyn's metropolitan pattern, discussed above, testifies. Movements in search of agricultural land persisted long after 1890. More homesteads were established after 1900 than before, for example, although there was again considerable variation by ethnic group (see Nugent 1992: 152).

The volume of transatlantic movement increased with the construction of railways on both sides of the Atlantic and with the introduction of progressively larger steamships (Nugent 1992). The increase and complexity of migration thus appear to have been a direct consequence of technological development. The relationship is, however, not nearly as simple as this might suggest, and gross national-level statistics are too coarse to allow an adequate assessment.

Transatlantic movements: of origins and cycles

Although there appear to have been widespread patterns of mobility in Europe, some groups moved more than others, and migration from some areas was much more prevalent than from others. Minorities figured prominently among emigrants from several parts of central and eastern Europe, for example, and one of the intriguing questions concerning emigration is whether it came from the poorer parts of European countries (Baines 1991: 31–2). The evidence is mixed, but perhaps the main point to be made is that cross-sectional data on emigration at any one period of time are unlikely to provide an accurate picture. Unless the migration is seen as a process evolving over time, spurious interpretations are likely to result. King (1993: 26), in a analysis of emigration from Italy, has shown how the emigration began in the mid- to late nineteenth century from the 'more developed, urbanized and accessible regions' of the north, with only later, in the early twentieth century, a shift towards mass emigration from the poorer south. A similar shift from more developed to less developed regions of origin appears to have occurred from Spain, Portugal and Greece, and King contrasts these long-developed flows with the much more recently developed flows from Turkey and former Yugoslavia to the Europe of the 1960s, which were still primarily from the more developed parts of these areas.

The type of migrant, too, varies over time, with the earlier migrants from urbanized areas tending to be better educated and more skilled than the later rural migrants. This pattern of development of international migration from southern Europe has clear parallels with that of internal migration in parts of the developing world for which I conceptualized migration systems diffusing down through an urban hierarchy from larger to smaller places and through a social hierarchy from wealthier to poorer groups (Skeldon 1990).

As migration was uneven across space, so too were there fluctuations across time, and there were several periods when emigration was of greater intensity: 1844–54, 1863–73, 1876–88 and 1898–1907, for example. There have been attempts to link these fluctuations with long cycles in capitalist development. North American and European building cycles were not in sequence and Thomas (1954) argued that, when the former was in the upswing and the latter in the downswing, transatlantic migration would dominate and that, when the cycles were in reverse, rural-to-urban migration within Europe would be most important. The evidence is conflicting as Thomas assumed that transatlantic migrants came mainly from rural areas and therefore had the choice to move either to North America or to urban areas within Britain depending upon job opportunities in the long cycle. When it was shown that most of the transatlantic emigrants from Britain came from urban areas, the presumed relationship between internal and international migration collapsed (Baines 1986). What proportion of the transatlantic urban migrants originally came from rural areas is not known, although it is likely that there was considerable step emigration from rural areas through urban areas and on to the Americas, a sequence that might have taken several years.

This is not to say that there is no relationship between economic cycles and migration but rather that the use of the state is too coarse a unit of analysis. There are likely to be strong intranational regional variations in business cycles depending upon the industrial mix of each region. Thus, regions within Britain may have been at different stages of a cycle depending on the product cycle of the industries in each area. This confuses the picture and makes it difficult to draw out specific relationships between national economic cycles and emigration. Inertia of population flow caused by chain migration and lag effects caused by delays in the diffusion of information concerning opportunities further complicate the situation. Despite these difficulties, relationships can be observed. Migration to Argentina, which accepted more migrants on a per capita basis than any other country in the Americas, certainly appears to have been in step with international business cycles (Adelman 1995). The boom years in the Argentinian economy in the 1880s, 1904–13 and 1919–24 were associated with peaks in immigration from Europe.

A marked relationship appears to exist between the volume of emigration and the growth of the population. Perhaps the only universal law of migration is that young adults have the highest propensity to move. When the number of births occurring twenty-five years before emigration in any particular year are graphed against emigration, a significant correlation is observed (Fig. 3.1). When the number of births increased, twenty-five years later an increase in the volume of emigration from Europe occurred – a relationship that appears for each of the four main waves of emigration (Thomas 1954; also Chesnais 1986: 171–3). There is a clear relationship between population supply and emigration that is much more exact than just a general association between the period of greatest demographic expansion in Europe's history in the second half of the nineteenth century and the Great Migration.

Labour deficit and enforced supply: the slave trade

Between the two main phases of European transatlantic migration came that other major flow across the Atlantic of slaves from west Africa, with the labour-deficit areas in the United States, the

Fig. 3.1   Natural increase and emigration, Europe, 1820–1915

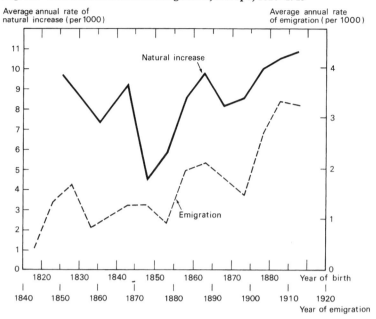

Average annual rate of
natural increase (per 1000)

Average annual rate
of emigration (per 1000)

Source: J. C. Chesnais (1992) *The demographic transition.* Oxford, Clarendon Press: 171.

Caribbean and Brazil being the principal destinations. The movements became significant from the late seventeenth century and, by 1870, between 9.6 million and 10.8 million slaves had landed alive in the Americas. Given a transatlantic death rate of between 10 and 20 per cent, and a death rate that could be as high as 50 per cent before the slaves even reached the ports in Africa, the numbers affected by the trade were very large indeed. The calculation of precise numbers is a complicated and controversial task based, as it is, on a multitude of far from complete or exact sources. The impact of slavery on the African continent is also difficult to assess. It was clearly variable and was felt in some areas and some ethnic groups more strongly than in others. The overall impact appears to have led to an absolute decline in the West African population.

These numbers are quite independent of those taken as slaves within Africa itself or exported to what is now West Asia or the Middle East – perhaps 4 million in the latter case (King 1996: 19). The evidence seems to suggest that, as the trade in slaves across the Atlantic was suppressed and contracted through the nineteenth century, so the internal and Arab trade intensified until it perhaps reached numbers equivalent to those of the Atlantic trade. As Lovejoy (1989) avers, in a comprehensive if surely not definitive review of the available data, slavery within Africa needs to be taken seriously by historians of both Africa and the Americas: its linkage with the Atlantic trade seems clear. The latter trade itself had a major influence on the demographic, economic and social development of Africa. For a more popular and wide-ranging review of the whole African 'diaspora', see Segal (1995). Slaves were crucial for the development of the export activities of the destination areas, primarily sugar and cotton. They were also important for the development of certain African kingdoms, for example, Asante, Dahomey and Benin, which controlled a significant proportion of the capture and transport of slaves. Despite the fact that many parts of Africa prospered and advanced politically and economically throughout this long period, other areas were adversely affected and it is difficult to balance any developments rationally against the cruelty, death and untold suffering brought by this trade in humanity. Although migration out of Africa has certainly not ceased, much of the continent at the time of the Atlantic slave trade was more closely linked to the international system than it was in later times, a theme which will be developed in greater detail in Chapter 7.

Economic and political retreat, and slowing migration

Emigration from Europe was disrupted by the First World War but was renewed in the 1920s at lower volumes than the pre-war movements. The great depression of the 1930s saw the effective end of this phase of European emigration and also marked a stabilization of urban growth in the core countries of Europe and North America following this economic crisis (Bairoch 1988: 33). Although further concentration of population in cities occurred after this period 'a less voluminous mobility surrounded urbanization after 1914 than in the past' (Moch 1992: 172). Fertility had dropped markedly throughout western and northern Europe from the late nineteenth century and, by the 1920s, the growth in those cohorts most likely to migrate slowed. By 1914, the transformation of the European world from agriculture to industry was essentially complete, with the continuing rural-to-urban migration much more permanent and with relatively stable populations in a series of very large industrial cities. The labour markets of that machine age were not so tolerant of high turnover and circulation back to rural activities. These too declined, with greater commercialization of agriculture and the drive for economies of scale.

The United States also continued a process of concentration

of population in urban areas through rural-to-urban movements, with the proportion of the total population in towns and cities increasing from 42 per cent in 1910 to 57 per cent in 1950. One of the trends to emerge from the 1920s was the migration of black Americans to the main industrial cities of the north. Although there had been attempts to settle slaves in the north, their high death rate rendered such endeavours uneconomic, whereas in the south their death rates were lower than those of whites (Fischer 1989: 52–3). Slaves had thus been concentrated in the rural south but, although emancipation dated from Lincoln's famous proclamation of 1863, few freed slaves moved to the rapidly growing urban areas of the north over the following sixty years. Partially, this reflected the sheer difficulties of overcoming prejudice and obtaining the resources to move: few would have had friends or relatives in the cities to foster chain migration and offer assistance at destinations. In addition, the large immigration from Europe towards those northern cities suppressed any demand for unskilled migration from other parts of the United States. Only after migration from Europe slowed after the First World War did opportunities open up for labour from internal sources, showing the relationship between international and internal migration in this case (see Long 1988: 145; also Nugent 1992: 155).

The period between 1914 and the end of the Second World War was not simply one of economic stagnation and consolidation of populations. It was, in the words of Hobsbawm (1994: 7), an 'age of catastrophe' in which not only was the process of globalization, or the previous century's extension of Europe overseas, reversed, but the liberal democratic institutions were almost extinguished in 'all but a fringe of Europe and parts of North America and Australasia, as fascism and its satellite authoritarian movements and regimes advanced.' Forced migration, slave labour and refugees came to characterize this period of European history as nations closed their doors and reduced their foreign labour in the face of the economic recession and the rising tide of nationalism. The lack of economic development and the emergence of repressive political structures engendered migrations of a very different type. Quite apart from the movement of millions of men during the various military campaigns of the First World War, hundreds of thousands of refugees were repatriated following the Armistice. Of some 2 million foreigners in Germany at the end of the war, only 174,000 remained by 1924 (Moch 1992: 165). These movements were dwarfed by the movements two decades later when Germany used over 7 million forced labourers. Millions of Jews were moved for internment and extermination and, in total, some 30 million people were displaced between 1939 and 1945 (Moch 1992: 168–9).

The era of development and migration

The triumph of liberal democracies and the task of rebuilding Europe ushered in a new period of prosperity, a 'golden age' – Hobsbawm again – but one that could just as easily be termed an 'age of development' or an 'age of developmentalism'. The idea of development long pre-dates this era (Cowen and Shenton 1996), but never before has the concept been so used, and abused, by politicians, policy-makers and academics. The notion that some form of rational planning could raise the level of living of populations, eradicate poverty and ensure an acceptable quality of life for all is fundamental to the modern use of the term 'development'. The basic and various concepts have been well reviewed in a companion volume in this series (Hettne 1995).

'Modern development' essentially began in Europe with the Marshall Plan to rehabilitate the devastation of war, funded by American grants with the aim of creating a strong western Europe as a bulwark against the communist systems to the east. By 1950, the majority of people in northern and western Europe lived in cities. The highest proportions of urban populations were found in Belgium and England, which were 91.5 and 84.2 per cent urban respectively at that time. Rural-to-urban migration persisted in areas such as France and Spain, where the levels of urbanization were lower, and the countryside continued the demographic decline begun in the inter-war period with rural depopulation becoming acute in more marginal areas (Fig. 3.2). Not that outmigration need necessarily lead directly to depopulation; net outmigration can be balanced against natural increase but as more and more young people, and particularly young women, spend longer and longer away from their communities, the reproductive age groups are eroded, leaving, in the final phases, only ageing and declining communities. Up to the 1970s, the dominant flow was, however, between and within urban areas, hardly surprising given the high proportion of population in that sector.

The prosperity of the post-Second World War developed countries was accompanied by the resumption of the process of globalization primarily through the expansion overseas of transnational corporations and international banks. International migration has once again become important, but there are major differences compared with that of the prior era of the Great Migration. The migration has become much more complex in terms of its origins and destinations and in terms of the types of migrant flows. Most importantly, while North America and Australasia have persisted as major destinations for international migration – more people were admitted into the United States during the decade of the 1980s than during any other decade in its

Fig. 3.2   Regions of net outmigration, Western Europe, 1961–1971

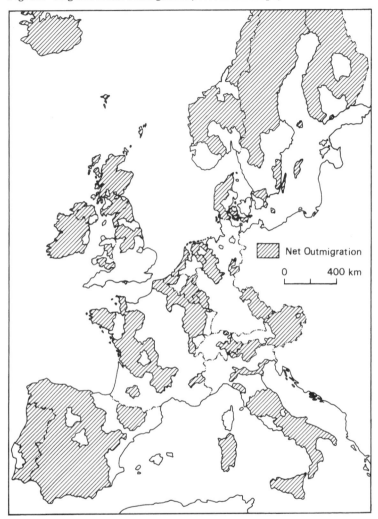

Net Outmigration

0   400 km

Source: J. Salt, H. Clout (eds) (1976) *Migration in post-war Europe.* Oxford, Oxford University Press: 37.

history, for example – the pattern of movement for Europe has been reversed: immigration has become more important than emigration. Emigration has certainly not ceased, but its composition has changed, and there is much European-based and American-based circulation.

The whole context of the migration has changed in two

significant ways. First, the developments in the field of transport allow rapid and relatively cheap movement between virtually any two major cities in the world within twenty-four hours. Regular commercial services by jet aircraft across the Atlantic date only from 1958, but it was the era of the wide-bodied aircraft, with the introduction of the Boeing 747 in 1970, which revolutionized personal mobility. The second change has been the change in fertility. After an initial increase immediately after the Second World War, the Baby Boom, there has been a sustained decline in fertility to the extent that, by the 1990s, there was not one country among the core developed countries with a total fertility rate (defined as the number of children a women will have borne by the end of her reproductive life) above the replacement level of 2.1. Most core countries have been significantly below that level for some considerable time. Low fertility and increasing international migration have also been accompanied by one other major population trend in the recent historical experience of the old core developed counties: urban deconcentration or counterurbanization. While no simple causal relationship exists between these three population trends, they are nevertheless interlinked.

The whole relationship between fertility decline and development is still not clear, and it is not direct. There are several countries or areas with much lower fertility than their levels of development would suggest, yet there are no highly developed economies with populations of high fertility. Sustained over a considerable period of time, below-replacement-level fertility will generate negative growth and ageing populations. When a stagnant or declining labour force growth is combined with the rising aspirations of the indigenous population as a result of their rising levels of education and personal affluence, a strong demand is created for the importation of labour. That labour is often to fill the so-called 3–D jobs, dangerous, demanding and dirty, that the indigenous population is no longer willing to do. However, as Böhning (1995) has correctly observed, these jobs are rarely dangerous, demanding or dirty – if they were they would almost certainly be highly paid and desirable to some – they are just low-paid, insecure and boring.

How the various old core countries respond to the demand for labour fuels one of the most controversial debates of our present time and one that will surely continue to tax politicians, policy-makers and the public into the next century. The issue is not simply one of a demand for labour; it also revolves around the supply. Given the low fertility throughout the old core areas, the supply must now come from outside the old core tier; the less developed countries of the world are being gradually drawn into an expanding international migration system centred around the wealthiest countries, and the wealthiest parts of these countries, the largest cities.

The changes in the international migration system over the last fifty years were outlined in Chapter 2. While the switch from a European emigration to North America and Australasia towards a movement from Asian and Latin American countries to North America and Australasia is clear, it is also obvious that not all of the less developed countries have been participating equally in the new migration. Rarely did the poorest countries participate significantly in the flows but ex-colonial ties, business ties and trading links were much more important in accounting for the observed pattern. It is not the poorest who migrate, so wealth differences are not necessarily important; what is more important is that peoples from different backgrounds and cultures are moving into what were once primarily European cultures in origin. The new migrants are, through modern systems of communication, often able to maintain close contact with their societies of origin, raising questions about national loyalty. As will be seen in Chapter 4, new extended families have emerged, with husbands, wives and children often living in different residences around the Pacific and family members virtually commuting at regular intervals between them.

The fears that some Americans have towards the new immigrants (see Brimelow 1995 for a recent example) echo those of a previous generation of white Protestant Americans from northern Europe towards the 'birds of passage', or short-term labourers, mainly from southern Catholic or eastern Europe. The hostility towards the migrant from a different background, particularly if he or she is seen not to make a permanent commitment to the host country, is certainly not a recent phenomenon (see Nugent 1992: 158–61 for a discussion of earlier antipathy towards circulators from Europe).

The debate about whether migrants make a positive or a negative net contribution to a country's development has once again moved towards the centre stage of public opinion and it becomes embroiled with issues of race and identity, as well as with welfare and employment. The main difference from the past, as emphasized above, is that while North America and Australasia have persisted as major destinations since the Second World War, Europe has emerged as a destination too, rather than the principal area of origin of migration. There have been, very generally, two different responses to the new immigration, despite variations among them, and these will be briefly discussed before we examine the trends in population redistribution within each part of the old core.

The new immigration: various patterns and citizenship

The two different responses to the new immigration apply to the original settler societies of North America and Australasia on the one hand and the 'new' destinations of Europe on the other. The former, countries created by immigration, continue to pursue policies that favour the acceptance of immigrants or settlers. The latter, all facing immigration as a new issue, have, with the exception of Britain, seen migrants more in terms of labourers than of settlers. From 1914, the United Kingdom allowed free access to all those born on British-administered soil. However, following marked migration from the Caribbean, and later from South Asian countries in the post-Second World War period, Britain began progressively to limit immigration from the 1960s onwards. The United Kingdom has also, for particular historical reasons, had a long and continuing tradition of 'free' migration from Ireland. With this exception, the European countries have viewed migrants as forming minorities rather than as contributing positively to nation-building, multicultural or otherwise. Unlike the settler societies, the European states have attempted to restrict, even stop, immigration, while the settler societies have, until the mid-1990s at least, generally maintained or increased their intake, albeit with some fluctuations.

Unlike the settler societies, where citizenship is open to all those who are accepted as immigrants and who go on to fulfil certain criteria within a relatively short period of time, the European states operate variable but generally stricter and more complex procedures towards the award of nationality. Some are relatively quick to award citizenship, like Sweden, while others are more restrictive, like Germany or Switzerland. Migration does not automatically lead to citizenship, and substantial proportions of migrants with permanent residence status in both settler and European states do not wish to change their nationality. For example, in the first half of the 1980s, only about 6 per cent of foreign residents in Germany, where the procedures are difficult, intended to apply for citizenship while, after ten years, the average take-up of citizenship was 25 per cent in the United States, with substantial variation between Asians at 48 per cent at the upper end, and Mexicans at 2 per cent at the lower end (Stalker 1994: 65–7).

The new immigration: the settler societies

Thus, over the last half of this century, we have seen a resurgence of migration to the settler societies but from very different sources from before. Officially, Australia, Canada, New Zealand and the United States have been pro-immigration, generally increasing their intake of settlers and asylum-seekers, though Australia in particular and Canada in the 1980s cut back their intake during periods of high domestic unemployment (Fig. 3.3). The declines in the second half of the 1970s and during the early to mid-1980s are clear in both Canada and Australia (see also Annexe Table 2). Interestingly, the high unemployment in Canada during the early 1990s did not bring a reduction in immigrant numbers and that country persisted with its five-year immigration plan. Australia, on the other hand, did respond to the economic slowdown by lowering its intake. Fluctuations for the United States have not been common in the post-Second World War period; steady growth continues. The larger numbers admitted between 1989 and 1991 are the result of the Immigration Reform and Control Act (IRCA) of 1986, which regularized the status of those who had arrived in previous years, rather than a sudden surge in movement in those specific years. For a discussion of the IRCA and its impact, see Calavita (1994). Although numbers of immigrants to all these countries are generally at their highest level since 1945, they are still below the levels seen in the early part of this century, when of course the total populations were much smaller and the relative impact of immigration was much larger. In the decade of the 1980s, the immigration to the United States of some 7.3 million added perhaps 3 per cent to the total population, whereas in the first decade of this century the 8.8 million represented at least 10 per cent of the total.

The new immigration: Europe as destination

Europe gradually ceased to be a major source of migrants from the late 1950s and was characterized by internal movements from south to north from Italy, then Spain, later Greece and Portugal, and then former Yugoslavia, to the booming economies of France and Germany in particular. Later, the source areas were extended to Turkey, North Africa and beyond (see King 1993b). The sequence of development of the migration, as originally identified by Böhning (Castles and Miller 1993), was given in Chapter 1, describing an initial movement of male labour into low-income,

Fig. 3.3 Immigration and unemployment: Australia, Canada and the United States, 1980–1993

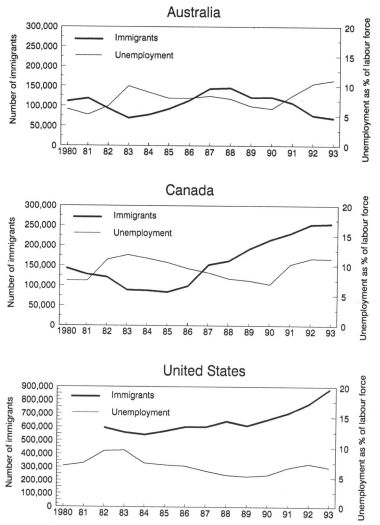

Source: Official government figures.

often short-term jobs, evolving over time to much more long-term, even permanent, family migration. Following the oil shock of 1973 and a series of recessions and persistent high unemployment, the European states sought to restrict immigration, initially with considerable success, and numbers dropped markedly. King (1993b),

Salt (1993) and Stalker (1994) provide some of the clearest and most systematic descriptions of recent trends in migration in Europe.

Despite the overall slowing of migration into Europe, the countries of southern Europe, beginning with Italy, then Spain and Greece, and finally Portugal, themselves emerged as destinations. The economic development of southern Europe from the early 1970s raised incomes and integrated it much more closely into an international division of labour, not only in the growth of light labour-intensive industries but also in services, not the least of which revolved around tourism. These economic changes, when combined with a general ease of access for outsiders compared with the countries of northern Europe, proximity to sources of supply of labour in North Africa and the Middle East and a slow growth of the indigenous labour force due to persistent low fertility, explain the turnaround in the migration systems in that area (King and Rybaczuk 1993).

The new immigration: common patterns

The differences described above between the settler societies and the European patterns, and even the differences within Europe itself, might appear to undermine any attempt to include them within the single development tier of 'old core' that could have any meaningful relationship with migration. Yet, these differences obscure important common patterns. The post-war development had spurred a demand for labour that could not be met from indigenous sources alone. The trend to extremely low fertility, with the slow growth of the labour force and rising expectations of local labour through improved education, reinforced the demand for foreign labour, particularly for low-paying service jobs. Immigration to all areas was coming from an increasingly complex set of countries of origin. There was a general emphasis upon family reunification, further emphasizing the links with home communities in all areas. Given restrictions or long delays in obtaining legal documentation, there were rising clandestine movements to all countries of the old core and increased movement of asylum-seekers or refugees. The latter became prominent in Europe after the disintegration of the Soviet Union from 1989 and the conflict in the former Yugoslavia. Germany, in particular, has been one of the main targets for those fleeing the new world disorder, receiving not only ethnic Germans from central and eastern Europe (excluding eastern Germany), some 1.2 million of whom arrived from 1988 to 1991, but also hundreds of thousands of asylum-seekers, with 438,191 applications in 1992 alone (SOPEMI 1994: 78). The United States, too, has been a major target for clandestine migration, mainly from Mexico, Central America and the Caribbean, but also from China and other parts of Asia; annual numbers may exceed 1 million, with the majority coming from Mexico. The United States accepts around 130,000 asylum-seekers per annum, with the most important sources being the republics of the former Soviet Union and countries of East Asia (SOPEMI 1994: 104).

Immigration has become one of the major policy concerns of all the old core countries, which explains the explosion of interest in the topic observed earlier in this book. These concerns mirror not only economic issues, such as the impact that immigrants might have on wage levels, unemployment and welfare benefits, but also the essential and sensitive political view that states have somehow lost control of their borders, and that the integrity, even the identity, of the nation is being eroded. The border areas, on the interface between different political and economic systems, have a special significance for migration from the point of view of development. This will be discussed in more detail in Chapter 5. The immigration of peoples holding different values and from different cultures has brought the Third World into the First World in popular interpretations of suitable vagueness. Paul Theroux (1995: 202) tells of an unknown traveller who pithily remarked, 'worlds can't meet worlds, but people can meet people' and the common-sense simplicity of this observation underlines just how outdated the terms Third World and First World have become.

It is difficult to separate fact from ideological conviction in such a highly charged debate, between those who favour immigration and those who wish to raise the exclusionist walls. There is indeed a problem of generalizing for all immigrants, thus undermining any overarching argument. For example, it is fashionable in both the media and the academic literature to examine the impact of 'Asian' immigrants, conveniently overlooking the variety of cultures and economic and political systems in an area that stretches from Iran to Japan. The conclusions that one reaches for highly educated, independent migrants are unlikely to apply to the unskilled family members that many migration policies, and particularly that of the United States, now favour. Even the skill levels for such a dynamic and entrepreneurial group as migrants from Hong Kong vary by destination; those going to Australia and Canada are much more highly qualified than those going to the United States (Skeldon 1994). Australia, Canada and New Zealand operate, to a much greater extent, a skill-based immigration policy, although the 1990 Immigration Act of the United States attempted to increase the proportion of independent skilled immigrants.

The impact of the new immigration

Immigration does not appear to affect aggregate unemployment rates adversely, even in times of recession. 'The weight of the evidence [for Australia and the United States] is that immigration has worked to lower unemployment' (Pope and Withers 1993: 735; see also Ackland and Williams 1992; and Tsokhas 1994). High unemployment may, however, act as a deterrent to immigration either through direct policy measures to slow migration taken by the core country, or through potential migrants staying in their country or choosing another destination. Overall, the lack of a major impact of immigration on local labour markets remains somewhat of a mystery, illustrating that we still do not understand 'the dynamic process through which natives respond to these supply shocks and reestablish labour market equilibrium' (Borjas 1994: 1700). Migrants do appear to be disproportionately represented amongst the lowest paid and in the types of jobs that indigenous workers shun, although this is an oversimplification as there appear to be two distinct groups of migrants. The first, and most visible, group is concentrated at the bottom end of the socioeconomic spectrum; the second consists of much more highly skilled professionals and entrepreneurs. The Indians are, by some systems of calculation, the most highly educated and affluent group in the United States, for example (Clarke, Peach and Vertovec 1990: 18), and the Chinese from Taiwan and Hong Kong have dominated the business migration programmes of Australia, Canada and New Zealand (Skeldon 1997c). Migrants such as these, it is argued, can only have a positive effect on a country's development; they create jobs, are unlikely to go on welfare and, by increasing competition, can benefit national consumers and international trade. While entrepreneurs can indeed create jobs, their success can be limited as they are guided into activities where the prospects are poor (Smart 1994). Where immigrants do create jobs, there may be exploitation of a labour force that is recruited from amongst members of the same migrant group. While these ethnic niches have been seen in a positive light, guaranteeing support and employment within the migrant group, there is obviously scope for exploitation, particularly in cases where migrants are being brought in illegally to be funnelled into ethnic businesses. Organized crime, bringing highly indebted Chinese into the sweatshops of New York, where they work as virtually bonded labour, is one of the more blatant examples of this type of exploitation. In the early 1990s, perhaps 100,000 Chinese were being brought into the United States illegally every year.

Ethnic niches and segmented labour markets are characteristic of most areas of high inmigration, whether internal or international, with the association and the mutual support group of rural migrants in cities in Latin America, Africa and Asia having their counterparts among the immigrants in the cities in North America, Australasia and Europe. Examples of exploitation within economic niches in the less developed parts of the world do not appear to be as marked as in the core countries, perhaps because, in the former, linkages between origin and destination are more close-knit and, in more familiar environments, fission and splitting of groups are possible in cases of disagreement and abuse. In the main destinations of the core countries, the ethnic niche is a protection against an often hostile host society, a source of mutual support and access to employment, but it can also be a prison, holding and exploiting new arrivals in an effort to maintain low costs in increasingly competitive environments. Migration to the old core countries, or to other destinations, can be a way to advancement, but it can also be a path to exploitation and abuse. See the debate between Bonacich (1993) and Waldinger (1993).

As migration from a particular source area increases through chain movements, the educational and skill qualifications of the migrants appear to decline. The United States, with its immigration policy heavily biased towards family reunification, has seen the skill and educational level of its immigrant intake decline markedly since the new immigration began in 1965 (Borjas 1990, Fry 1996). Some groups are indeed highly educated, but others are much less so and may consume more than they produce. Even those who belong to the upper echelons may avail themselves of state benefits in certain ways as, for example, in the case of wealthy Asian immigrants to Australia who use government grants to educate children who later return to their home areas without repaying the loans (Birrell and Dobson 1994). Such short-term costs have to be balanced against more long-term gains of increasing numbers of Asians who stay, pay their taxes and foster growing trade links with their home countries, which are currently the world's fastest growing economies.

However, in the context of 15 million unemployed in Europe, arguments to allow still more immigrants to enter are difficult to sustain politically. Existing labour-force participation rates are the lowest of all the major trading blocs (Coleman 1992: 435) and the hidden labour supply will more than compensate for any decline in the size of the labour force owing to the changing age composition and ageing of the populations over the medium term. Increasing immigration, it is argued, may inhibit the incorporation of indigenous women into employment and thus the improvement in their status. Until the existing migrant groups have been integrated more successfully into national economies and removed from their marginal position in society, it is pointless to encourage further immigration. Such 'fortress Europe' thinking ignores the very real contributions that migrants can and do make towards relieving unemployment and represents a different approach from the official and, almost certainly, majority views in the settler societies, although there, too, there are similar and vociferous minority views against further immigration.

It is probably impossible to come to a truly objective cost–benefit assessment of so complex a phenomenon as the impact of immigration when many of the benefits and costs lie in an unpredictable future. It is even more difficult to be objective in such an emotionally charged debate as that between those who are pro-immigration and those who are anti–immigration. Nevertheless, the balance of the research carried out on this topic is that immigration brings more benefits than costs, that migrants create more wealth than they consume and that they have a minimal impact on native wage levels and increase labour market flexibility (Wooden 1994, Friedberg and Hunt 1995, Zimmermann 1995). In addition, the benefits that immigrants bring to native populations increase with the skill level of those admitted (Borjas 1995).

The urban orientation of the new immigration

One similarity between the new immigration to Europe and to the settler societies of the old core is that the migration is primarily urban-oriented and is concentrated towards a few large cities in each case – the global cities. This is not to say that all migrants go to those cities. As Portes and Rumbaut (1990: 34) have observed, there are two contradictory outcomes of the immigration to the United States: a concentration on a few large metropolitan areas but also a dispersal of a minority of immigrants throughout the country. In 1993, some 70 per cent of the 904,292 immigrants admitted to the United States went to just six states, California, New York, Florida, Texas, New Jersey and Illinois (INS 1994), virtually the same proportion as in 1987 (Portes and Rumbaut 1990: 34–5; see also Massey 1995), with some 40 per cent of the new immigrant population being found in New York and Los Angeles. Similar patterns of concentration in the largest cities have been observed for Canada and Australia, and for European countries (King 1993b: 26).

A bipolar division into two very different groups of workers is seen as a characteristic of global cities and is a direct function of migration. On one side, there is a wealthy, highly mobile group of professionals, often in financial services, while on the other there is a group of poorly paid workers who staff the basic services required by the professional group (Sassen 1991). Both groups include high proportions of migrants, although attention tends to be focused on the latter, which is seen to be made up of poor immigrants from the less developed world, often women and those with a poor command of English, who are in casual service activities. They have none of the health or pension rights that the professional groups receive. The growth in the number of poor immigrants has been seen to promote the expansion of the professional classes as the 'immigration lowers the cost of recruiting and maintaining this "new labour aristocracy"' (Waldinger 1992: 99) by subsidizing the demand for an increasingly complex range of services. The reality, as Waldinger (1992) has shown, is likely to be more complex, with the immigrants more widely spread through a range of skills in service activities. More important, rather than immigration driving the growth of the professional group, and of the service sector itself, the immigrants are responding to a demand created by a decreasing supply of indigenous labour in the world cities. This decreasing supply is a function not only of declining fertility but also of movements away from the largest cities. Thus, there are links here with the recent patterns of internal migration in the old core tier.

Counterurbanization

One of the unexpected trends in the countries of the old core tier in the 1970s was a reversal of the traditional migration to the largest cities in a move towards population deconcentration, counterurbanization or urban turnaround. This deconcentration, together with the acceleration of the ageing of the populations consequent upon low fertility and changes in household composition, due to low fertility but also to divorce and other changes in family formation, make up the principal demographic trends of the more developed countries (Champion 1992). The trend towards a migration away from the largest cities first occurred in Britain from the beginning of the 1960s; it then peaked in the 1970s and has since lost its momentum (Champion 1994: 1504). An interesting aspect of this migration is that it may have occurred, in Britain at least, in a series of steps by groups of people moving sequentially in short moves down the urban hierarchy. This 'cascade effect' (Champion 1995) is a reversal of Ravenstein's system of rural-to-urban migration observed in the nineteenth century.

In the United States, as in so many other old core countries, the phenomenon of counterurbanization characterized the 1970s, but has slowed there too, and there has been a renewed growth of metropolitan areas in the 1980s and 1990s (Frey 1993). These shifts in internal movement are associated with changes in the international economic system which have seen the decline of labour-intensive manufacturing in the face of competition from Asia and elsewhere. The resulting economic restructuring is towards light industries (which require physical space and thus locate away from the traditional congested centres) and towards services. The most recent pattern of migration is not simply away from the largest metropolitan centres towards new areas of manufacturing; it is complicated by the immigration which, as discussed earlier, is to the largest metropolitan centres. A pattern can be discerned in which the states with the highest gains from international movement are also among those of high net outmigration (Frey 1995). The outmigration from these states tends to be of whites with low income and a low level of education, while elites still circulate between and to these centres. There appear to be no net losses of elites in states of high immigration, reflecting the bipolar structure of the large and global cities discussed earlier with elite indigenous groups and poor immigrant groups. The flight of lower-middle-class and working-class whites from the centre and suburbs of the largest cities towards smaller centres in the south and west has left opportunities for the immigrant group which are reinforced through chain migration and ethnic niche economies. This is leading to what Frey (1995), perhaps rather dramatically, describes as the demographic 'balkanization' of the United States.

Whether these patterns are being replicated in other old core countries awaits further research on the direction of internal movements and their relationship with immigration, which have been so clearly demonstrated for the United States. Even in the United States, there are clear variations by area, by skill group and by time period. In addition, different interpretations can result from different definitions of 'migrant' and from the ways in which the time periods are bounded, showing, to re-echo the words of Borjas cited earlier, that we still understand very little of the complex interrelations between internal and international migration and labour markets (see also Barff, Ellis and Reibel 1995).

Of elites, emigrants and retirees

The movement of the highly educated towards and between both large and small metropolitan areas remains a characteristic of old core internal movements. Many are company transferees or people moving to corporate headquarters. With growing globalization through multinational corporations, these internal transfers merge virtually seamlessly into movements across borders but within the network of the firm. Findlay (1995) terms these migrants an 'invisible phenomenon' because they are not perceived to be a social threat or to pose a burden for any individual country. They will normally be short-term transferees, for a few years at most, and are the responsibility of the company, not the country. Given the agglomeration of corporate headquarters in certain large cities, there is a concentration of elite migrants in these cities. Not all elite migrants are transferees, of course, but these cities act as magnets to attract the highly skilled and ambitious in a global market in high-level services, where salaries and cultural and social facilities are amongst the highest on offer (see Beaverstock 1994).

The international movement of these elites draws attention to the fact that, although the old core tier is now made up mainly of countries of net immigration, emigration too has either persisted or become important. The impact of emigration has perhaps been greatest in the smallest of the settler countries, New Zealand, which has experienced net emigration since the mid-1970s. This has consisted mainly of trans-Tasman migration towards Australia of white settlers and native New Zealanders, who are only partially replaced by migrants from the Pacific islands (see Chapter 7). However, emigration from Australia, the United States, Canada, Britain and other countries is also an important and little recognized part of global international migration systems. Some 2 million emigrants were estimated to be living outside the United States in mid-1988 (excluding those affiliated to the United States government, that is, members of the armed forces and other official employees), of whom about half had left in the period from 1980 to 1988 (Woodrow 1990). Thus, emigration accounted for about one third of the immigration over the same period. Of the most recent emigrants, about 46 per cent were American-born; the rest had been immigrants. While some 40 per cent moved to other places in North and Central America, the emigrant migration fields, like the immigration fields, extended virtually worldwide. These modern flows echo Ravenstein's laws: each current of migration produces a counter current. Canada and Australia show approximately similar levels of emigration, reviewed by Hugo (1994a), each with significant numbers of immigrants returning to their birthplaces and the Australian-born or Canadian-born being highly skilled and educated people: the circulation of a professional elite around global and regional cities. This topic will be further considered in Chapter 4.

Among the emigrants from the old core countries are significant numbers of retirees. The issue of migration and the elderly has become important as the old core populations have aged. There are two different aspects. The first is where the ageing of areas is often a consequence of migration, with younger people, with higher mobility, moving out, and leaving the elderly behind. The second is when the elderly themselves move and, when they do, they tend to migrate over relatively short distances. Nevertheless, more long-distance movement of the elderly to destinations that are seen to have specialized amenities and an equable climate is of growing importance.

Florida, southern California, Arizona and Texas, the Gold Coast of southern Queensland, Australia, and the south coast resorts in England are all well-known examples of destinations for the internal migration of the elderly (see Rogers 1992), but there are also significant international movements. Within Europe, there are movements from northern to southern Europe, particularly from the United Kingdom to southern Spain. Much of the emigration just discussed from the old core countries consists of immigrants returning home after retiring from their life's work. These growing movements will have important consequences for the development of places of origin and destination of these elderly migrants. For example, every year, Australia pays more than A$22 million to retired British migrants alone, who have since left Australia. In the case of Britain itself, 700,000 pensioners, or about 7.5 per cent of the total, live overseas on state pensions. Pensions form an important source of remittances to some societies, although those societies themselves may have to bear much of the cost of the provision of services, training of specialist nurses, doctors and so on, for elderly people. The cost to the country where the migrant spent most of his or her working life, of sending overseas large sums of money, which the current labour force is essentially generating, may become a future political issue as the number of retirees increases relative to the number of workers. These aspects of migration are likely to be of growing concern as populations age.

Migration for recreation

The final type of mobility that is generated from the old core countries that has a profound impact on development is short-term, but often long-distance, migration for recreation. Tourism has become one of the world's leading industries, and the affluent groups of North America, Europe and Australasia (and increasingly East Asia) with leisure time are the principal consumers. Most tourism is within or directed towards the old core countries themselves – some 73 per cent of tourist arrivals in 1994, for example, were either towards or between the countries of Europe or North America – but the relative development impact can be greatest in smaller and poorer areas. Consequently, this important topic will be examined in Chapter 7.

Discussion

This chapter has attempted to review the evolution of population movements over the last 400 years between and within what are now the most developed and highly urbanized countries in Europe, North America and Australasia. Only the most salient aspects of this vast topic could be raised. What is becoming apparent, and should become clearer from the discussions in subsequent chapters, is that, despite all the differences in the historical experience of the countries in the old core, there is a growing convergence of migration experience: convergence to the extent that all old core countries are the destinations of increasing immigration from poorer areas and most are now countries of net immigration; all have experienced a phase of concentration of population followed by deconcentration, which has since waned; and all are participating in transnational circuits of the circulation of the highly skilled.

Faced with the increasingly complex range of migrations, the most developed countries have begun to show a convergence in the types of immigration policies that the various governments have tried to implement and a gap has been growing between the goals and the outcomes of these same policies (Cornelius, Martin and Hollifield 1994). Thus, despite the increasing volume and complexity of migration to, and within, the old core tier, there is growing resistance towards migration, with political pressure to adopt more restrictive policies to control and limit population movements. The idea of an inevitable convergence is also central to the French sociologist Dominique Schapper's writings on the integration of immigrants in Europe, although see the commentary by Mandel (1995).

The idea of convergence must not be pushed too far, however, as there are still important differences within the old core tier. Rates of residential mobility vary markedly between the countries, with much higher rates for the settler societies in North America and Australasia than for Europe, or for Japan for that matter (Long 1991). Given that there has been a convergence in family structure formation and dissolution, and in lifestyle, in the most developed countries, variations in mobility are unlikely to be explained by any differences in the life cycle. Those factors that influence the housing market and the degree of government involvement therein appear to be more important. Where the markets are relatively open and where finance is relatively easy to obtain, residential mobility appears to be much greater. Thus, government intervention in housing and labour markets in more developed economies may indirectly be more effective in affecting mobility than direct government intervention in trying to limit mobility in less developed economies.

While there may be convergence in the general patterns of migration among the old core areas, even if considerable variation remains in the volume of movement, a blurring of differences is also occurring between these countries and areas within the so-called developing world. The recent movement of peoples from a variety of origins has changed the ethnic and cultural mix of old core destination areas. These destinations are primarily urban and the emergence of increasingly polarized social structures between professionals on the one hand and low-paid service workers on the other, as well as the existence of ethnic niches and segmented labour markets, is familiar to those who have worked in the less developed parts of the world. The blurring is facilitated by the human circulation within the networks established between origins and destinations, and it is these networks that have done so much to transform, not only the destinations, but also the origins. It is to those regions that we now turn, beginning with those that have reached levels of development that rival those of the old core countries. These areas lie in East Asia and make up the new core development tier.

Chapter 4

The new core

Of Asian models of development

In the aftermath of the Second World War, most of East Asia, like Europe, lay in ruins. Refugees and poverty characterized the area. In the early 1950s, the Korean War brought renewed conflict and population shifts. Yet, out of the devastation of war have come the most dynamic economies in the world, producing not only levels of prosperity that rival, in some cases even surpass, those of the old core just discussed, but also challenging theories of development. Out of the periphery can indeed evolve prosperous economies that have achieved balanced and self-sustaining development. So has emerged the East Asian model of economic development: the World Bank's 'East Asian miracle' (World Bank 1993). The most developed of the new core countries is Japan, but to that country must now be added the 'newly industrialized economies' (formerly known as the 'newly industrializing economies') of South Korea, Taiwan, Hong Kong and Singapore, the NIEs. Their development has continued at such a pace and their economies have been so restructured that we can now validly speak of the 'newly dein-dustrializing economies', as they have moved significantly from labour-intensive manufacturing towards service economies.

Japan, Hong Kong and Singapore have been classified by the World Bank as 'high-income' economies, and South Korea is at the top of the list, after Portugal, of the 'upper-middle-income' economies. Taiwan, which is not part of the World Bank system, has an estimated GNP per capita about 50 per cent higher than that of South Korea and a third lower than that of Hong Kong (*Asia Yearbook*, 1995). All these areas are intensely urbanized, with Japan and South Korea being classified as having 77 and 78 per cent respectively of their populations living in the urban sector in 1993, while Taiwan has 94.3 per cent of its population living in localities of 20,000 people or more (ROC 1990). Hong Kong and Singapore are essentially city states. The populations of all these areas have very low levels of fertility, amongst the lowest in the world and hence, like Europe, will age quickly. The governments in all these areas are in effective control of their territories and all can be classified as 'strong' states. They may not have the long tradition of popular participation of western democratic systems, but most can be considered democracies of a sort, or at least have energetically emerging movements for popular participation. Even where the state may appear more authoritarian, there are evolving systems for the rule of law to guarantee individual rights and freedoms and to prevent abuse in high office.

While the discussion of Hong Kong and Singapore in the category of the new core tier is logical under most developmental criteria, their inclusion does amply illustrate the difficulties of drawing boundaries that were discussed in Chapter 2. As two relatively small cities, they have extended their economic activities into hinterlands that are part of other sovereign states, even if, in the case of Hong Kong, it will shortly be part of that other state. Thus, it might be more logical to discuss these two cities in Chapter 5, the rapidly expanding core. Of course, to some extent, this issue applies to all countries and regions within the context of a growing international division of labour but, in the case of Hong Kong and Singapore, the geographical contiguity means that the boundary bisects an integrated economic region. While discussion in Chapter 5 will clearly have to include these and other rapidly developing areas, Hong Kong and Singapore will be introduced in the new core, as they are clearly a part of the emerging economic core in Asia.

The emergence of the new core areas has given rise to the idea that there is an Asian model of development distinct from that of the old core, one that depends heavily on state intervention and on Asian values, which, in turn, are seen to be derived from Confucianism (see, for example, Redding 1990). The formative role that the state played in early European industrialization and development (Weiss and Hobson 1995) is often conveniently forgotten. And the attribution of a role to Confucianism is reminiscent of the perceived importance of a Protestant ethic in northern and western European development. There are important differences between the European and Asian development experiences, but more comparative work is required before we can say that there is a truly separate Asian model. Many of the so-called Asian values have also had their counterparts in other cultures at various times. A discussion of the major reasons why the East Asian region should have emerged as the region of most dynamic growth in the world and whether this is an entirely different model from that of the west lies beyond the scope of this book. Here I will examine the changing migration patterns and venture comparisons with those just discussed for the old core tier.

The historical background of migration in East Asia

As in the case of the old core tier, there has been considerable variation in the experiences of areas within the new core. All, however, were relatively late developers compared with the pioneering areas of the old core. Although Japan had evolved an intensive urban network under the Tokugawa regime (1603–1867), with classes divorced from agricultural production, it was not until the mid-nineteenth century, when it was opened to the outside world, that its technological backwardness became clear to local leaders. Japan was as urbanized as Germany and the United States, for example, in the mid-nineteenth century (Table 1, page 63). The resultant stresses in the society brought about the end of the Tokugawa Shogunate with the Meiji restoration of the emperor from 1868 and a drive to modernize the country based upon foreign technologies. For an incisive discussion of Japan's economic development, see Francks (1992). Taiwan and South Korea remained essentially agrarian societies and became colonies of an expansionary Japan in 1895 and 1910 respectively to supply raw materials to the rapidly industrializing heart of this new core tier.

As in Europe, pre-industrial forms of population mobility were widespread, again disproving the idea that peasants were bound to the land. There were legal restrictions on movements, but temporary labour migration (*dekasegi*) of both men and women over considerable distances was common in Tokugawa Japan (Hayami 1973, Hanley 1973). These movements can be traced back many centuries into the pre-Tokugawa period (Farris 1985). Taiwan was a settler society for migration from China, beginning on a large scale during the Dutch occupation in the seventeenth century and accelerating after the island became part of Fujian Province in 1683. Much of this appears to have been seasonal migration from Guangdong Province and more permanent migration from Fujian Province. By the early nineteenth century some 2 million Chinese settlers were recorded, with almost 3 million a century later (Ho 1978: 11). They opened up the island to specialist agricultural production, especially sugar and tea, as well as rice, extending the cultivated area and stimulating trade.

During the Japanese colonial occupation of Taiwan from 1895 to 1944, migration from China slowed and was again mainly circular, while that from Japan increased. The former was discouraged, while the latter was encouraged. The Japanese not only included colonial administrators but also farmers, and the majority moved as nuclear families, while the Chinese migration, being circular, was overwhelmingly male. By 1943, the Japanese population had reached 384,000, up from 71,000 in 1906, when systematic records began, but the situation changed completely after the Second World War. Information on these movements is given in Chen and Liu (1996).

Singapore and Hong Kong were established as outposts of the British colonial empire in 1819 and 1841 respectively but, like Taiwan, their development was heavily dependent upon migration from China. Hong Kong was one of the principal ports from which people left China for other destinations, initially to North America and Australasia in the 1850s to 1880s in the gold rushes, and later to Southeast Asia when the former destinations began to exclude Asians (Chapter 3). By 1939, some 6 million persons, some of whom would have moved more than once, had passed through Hong Kong on their way overseas (Sinn 1995). This migration laid the basis for a virtual global network of Chinese communities that was important for the development of the much later movement towards the end of the twentieth century (Skeldon 1996). Hong Kong itself was also a destination for migration from China and its population rose from a few thousand in 1841 to reach 181,437 in 1886 and 456,739 in 1911 (with only about one quarter of the increase due to the physical expansion of the territory). Singapore similarly grew from a very small population – in 1824 there were 3317 Chinese out of a total population of 10,683 – to reach 303,321 in 1911, almost three-quarters of whom were Chinese (Ee 1961). The populations of Hong Kong and Singapore were heavily biased towards males with, as in the case of Taiwan, much circulation back to China in a classic sojourner pattern.

As the Chinese were emigrating upon the expansion of western influence into Asia, so too were the Japanese, though not in such large numbers: 'much less than one million' left between 1815 and 1932 (Shimpo 1995). The United States, especially Hawaii, was the initial and principal destination, although Peru and Brazil became significant destinations in the early twentieth century as the United States progressively closed its doors to Asian immigration. In the United States, the Japanese became highly successful in certain activities: they came to control some 80 per cent of the vegetable market in Los Angeles and they were the only immigrants wealthy enough to drive cars in Honolulu (cited in Howe 1996: 392). Their very success, though, engendered the resentment that led to their essential exclusion following the Immigration Act of 1924. The proportion of Japanese aged 15–59 years outside their homeland increased, however, from 2.6 per cent in 1920 to 3.2 per cent in 1930 to 5.6 per cent in 1940 (Taeuber, cited in Davis 1963: 347), as the Japanese turned to other destinations, mainly in Manchuria and Taiwan. Although only 33,070 Japanese had gone to Peru by 1940, many of whom subsequently returned to Japan (Shimpo 1995: 49), this quite limited migration has had a profound impact on the political and economic development of the country. The present president, Alberto Fujimori, the descendant of Japanese migrants, has played a significant role in guiding Peru out of the economic and political morass into which it had fallen in the 1980s during years of populist leadership and continual guerrilla activity. This example highlights the significant way in which a dynamic migrant group can influence development in destination areas.

The drive towards rapid modern development

As emigration was increasing, Japan was undergoing its long transformation towards an urban industrial society. As the textile industry spread in the late Tokugawa period, the contract labour (*dekasegi*) became more urban-oriented and progressively more long-term (Hayami 1973). By the end of the century, 'a definite rural-to-urban migration had set in' (Kawabe 1984: 123), although Japan probably had only about 15 per cent of its population in urban areas by the end of the first decade of the twentieth century (Kuroda 1986) and rural-to-urban migration had relatively little impact on villages until after the Second World War. By then, eldest sons, too, were participating as much as younger sons in the movement to towns, undermining the importance of kinship and ultimately eroding the reproductive capacity of the rural sector. The resultant rural transformation, however, need not be seen as unbalanced and distorted but as something new and constructive, even in the absence of the population (Knight 1994). Japan, unlike the other areas in the new core but like the old core countries in northern and western Europe, had experienced relatively low fertility and population growth during the lead-up to its modernization. Fertility increased slightly in the early Meiji period and population growth peaked, probably during the second decade of the twentieth century at over 1 per cent per annum, after which there was a sustained decline in fertility (Ogawa and Suits 1981). Mortality had begun to decline much earlier and did so over a long and gradual period. Although Japan was a 'late starter' in economic development compared with northern European and North American old core countries, its process towards industrialization was also very gradual and cumulative, and it took a long time to achieve rapid economic growth, which really only occurred after 1954 (Ohkawa 1983).

Even accepting that population growth in Japan was low in

comparison with less developed countries today, it was rapid compared with the almost two centuries of stagnation that had preceded 1868, and it was seen to be a problem (Howe 1996: 388). The closing of North American destinations and the perceived need to find outlets for the 'excess population' further encouraged Japanese expansionism into China, particularly in the face of a northward migration of almost 4 million Chinese, mainly from Shandong and Hebei, into Manchuria in the late 1920s, even if about half of these returned relatively quickly (Howe 1996: 390–6). In total, there was a net migration of some 8 million migrants into Manchuria during the first half of this century (Gottschang 1982).

Japan is generally seen as an essentially homogeneous society that has never resorted to the use of immigrant labour to foster its development. As far as the post-Second World War period up to the mid-1980s is concerned, this impression has more than just an element of truth. However, it must not be forgotten that there was a net immigration of some 400,000 Koreans during the 1920s and that the Korean community grew from less than 4000 in 1915 to over 1 million in 1939 and to over 2 million in 1945 (Weiner 1994: 198; also Lee and de Vos 1981). Koreans thus provided cheap labour for Japan's heavy industries and, although some 1.4 million were repatriated after 1945 (Kim and Sloboda 1981), large numbers remained as a permanent part of the labour force. By far the best and most detailed account of Korean migration to Japan and Japanese attitudes towards Koreans is Weiner (1994). Koreans represent a relatively small proportion of the total population of Japan – about 600,000 out of a population of 125 million in the 1990s – and although their role in their host country's development was hardly insignificant, their relative impact was probably greatest on the northern part of their homeland. About half of the Korean community in Japan came from what is now North Korea and their remittances, estimated at between US$600 million and US$1 billion per annum, were important in sustaining a near-bankrupt economic system in recent years (Chung 1995). The whole issue of remittances and development is considered in Chapter 6.

The development of the other four parts of the new core tier has been different from that of Japan. The four little tigers of Asia, South Korea, Taiwan, Hong Kong and Singapore, have shown some of the most rapid and sustained economic growth rates in history. Compared with Japan, all four were 'late starters'. In 1950, Hong Kong and Singapore were sleepy entrepôts on the periphery of a dying empire and Taiwan and South Korea were agrarian economies, even if Japanese colonialism had laid the basis of a sound infrastructure. Only 18.3 per cent of South Korea's population was classified as living in urban places in 1949 and the proportion living in Taiwan's five largest cities in 1950 was virtually identical (ESCAP 1975, Speare 1974, Ho 1978). Compared with Japan, fertility and population growth were significantly higher during the early phases of their rapid industrialization. All had total fertility rates in excess of 6 around 1960. Again, fertility appeared to rise during the initial phases of development and both economic development and fertility decline were much more rapid when compared with Japan. Taiwan, for example, was able to absorb its surplus agricultural labour in just twenty years (1952–72) compared with fifty years in Japan (1870–1920) (Ohkawa 1983: 56). By the 1980s, fertility was below replacement level in all four areas of the new core, Hong Kong, South Korea, Singapore and Taiwan, a decline that had taken just over twenty years. Japan achieved below-replacement-level fertility by about 1960, the decline having been initiated some fifty years before.

Migration in East Asia: a refugee mentality?

International migration has figured much more prominently in the recent history of the four outlying parts of the new core compared with Japan. As seen earlier, Taiwan, Hong Kong and Singapore demographically were largely a product of migration from China, and migration continued to be significant in the immediate post-Second World War period. Hong Kong had experienced dramatic fluctuations in its population in the late 1930s and 1940s. First swollen by some three-quarters of a million people fleeing the Sino-Japanese war, it was then depleted by almost 1 million during the Japanese occupation of 1941–5. With the re-establishment of British rule and the later triumph of communist forces in China, well over 1 million re-entered the colony.

The fall of Japan and the later defeat of the nationalist armies in China also brought major changes to the population of Taiwan. Japanese military, government and business colonial personnel, who represented over 5 per cent of the population, or over a quarter of a million people, left after 1945 (Copper 1990: 25). Between 1946 and 1950, somewhere between 1 and 2 million Chinese fled to Taiwan (Pannell and Ma 1983). Many were military personnel, but there were also large numbers of civilians including 'a small group of managers, technicians and entrepreneurs [whose] arrival enabled Taiwan to partially bridge the human resource gap that developed after the Japanese departed' (Ho 1978: 105).

Singapore, too, was profoundly affected by migration, though its population in the post-Second World War era appears to have been relatively stable compared with Hong Kong and Taiwan. At the outset of the war, Singapore's population was about 800,000, to which was added about half a million refugees from the Malayan peninsula who had fled the advancing Japanese (Thio 1991). Many of the latter returned home during the occupation of the city by the Japanese but, in the post-war period up to 1957, migration from the peninsula was significant, adding about 1 per cent per annum to the total population (Saw 1991).

South Korea was also a land of migration in the immediate post-1945 period. Some 2.5 million entered what is now South Korea between 1945 and 1949: as mentioned earlier, there were 1.4 million repatriated Koreans from Japan, but also another 620,000 from Manchuria and 460,000 from North Korea (Kim and Sloboda 1981). There are estimates (reviewed in ESCAP 1975) of another 1.6 million people moving from North to South Korea over this same period. In 1945, South Korea had a population of about 26 million, to which were added another 3.5 million in a matter of four years. The Korean war, from 1950 to 1953, created enormous destruction and hundreds of thousands of refugees, with almost two thirds of a million arriving from the north and a quarter of a million moving to the north, while countless others moved within each of the two parts of the country and principally to the largest cities of Seoul and Pusan. Hence, South Korea, like Hong Kong and Taiwan, was characterized by mass migrations immediately before embarking upon its course of most rapid economic development.

Migration certainly did not cause the rapid development of the East Asian economies and does not rank with the macro-political and economic factors that are so often cited. The role of public policy in facilitating free markets and of American food and technological aid in the support of systems on the interface of communism and capitalism were clearly critical in this process. Migration nevertheless served to concentrate in particular locations small groups of highly educated or entrepreneurial people who were able to take advantage of the opportunities offered. This minority among the minority of migrants, in situations of insecurity and uncertainty, produced creative tensions that drove a search for material progress and security. This 'refugee mentality actually creates a spirit of enterprise and engenders economic dynamism' (Wong 1992: 931). Obviously, though, not all refugees are either blessed with such qualities or put in a situation where such qualities might flourish.

A survey of the refugees who entered Hong Kong between 1945 and 1952 revealed that a quarter of the adults had been businessmen, professionals, intellectuals or army and police officers (Hambro 1955). The majority of those going to Taiwan and from Japan to South Korea had had urban experience with, as mentioned above, a minority of the former being professionals, businessmen and administrators. The issue of current refugees will be examined in more detail in Chapter 7, but the example of East Asia in the immediate post-war period illustrates the difficulty of distinguishing refugees from 'economic migrants': fully 37.2 per cent of the heads of refugee household in Hong Kong, for example, stated that they had come for purely economic reasons. These examples also emphasize the need to incorporate refugees into any integrated framework of migration and development.

The refugees gave a tremendous boost to urban growth in the immediate post-war period in Hong Kong, Singapore, Taiwan and South Korea, and urban growth continued in the 1950s and 1960s with massive rural-to-urban migration in Japan, South Korea and Taiwan. As Hong Kong and Singapore are essentially city states, their borders became more effective barriers to movement at certain times. In the case of Hong Kong, however, the border has been very porous, with continuous cross-movement, even if there were pronounced waves of migration and periods of lull. Emigration was virtually prohibited during the Maoist era in China, and the only major exodus was to Hong Kong following the famine conditions in China of 1959–61.

Labour surpluses, labour deficits and migration transitions

The maintenance of labour-surplus urban economies was therefore an important part of supporting competitive industrial enterprises at critical stages in the development of the peripheral parts of the new core tier. The opportunities provided by the Korean War stimulated employment in Japan, Hong Kong and Taiwan, and urbanization accelerated. In Japan, there was a clear trend from a dominance of short-distance intra-prefectural movements in the 1950s towards more long-distance inter-prefectural movements in the 1960s, mainly to metropolitan destinations as the overall mobility rate also increased (Kawabe 1984). This pattern was clearly related to the economic growth in Japan over this period, with prefectural disparities increasing and migration accelerating in consequence (Ogawa 1986). From being around 37 per cent urban in 1950, a figure similar to the immediate pre-war estimates, the proportion reached 72 per cent within twenty years. The peak year for transfer of population from non-metropolitan to metropolitan areas in Japan was 1961. In Taiwan, urban growth was fastest in the 1940s and 1950s (Speare 1974), when natural increase and internal migration were both high. In South Korea, the peak rates of urban increase were somewhat later, in the late 1960s and early 1970s, with the country only 17.2 per cent urban in 1949, rising to 48 per cent by 1975 and 78 per cent by 1993. The rate of growth of the largest cities in all these areas was fastest during the early period of urbanization.

The shift from rural to urban areas in East Asia has been accompanied by a degree of integration in labour markets that was much closer than in most developing economies. Fields (1994: 11) argues that there were two major phases in the development of these labour markets: a phase of increasing employment at essentially constant wages followed by rapidly rising real wages with generally full employment, which reflected a transition from labour surplus to labour deficit. The turning-point from labour surplus to labour deficit for Japan, as measured by the point at which real wages started a sustained increase, occurred shortly after the Second World War, some eighty years after it began to industrialize. Taiwan reached a turning-point around 1970, with South Korea becoming labour-deficit after 1975. Malaysia and Thailand, to be considered in Chapter 5, probably passed their turning-points in the mid-1980s (data reviewed in Manning 1995; see the essays in Abella (1994b) for similar data). The driving force in this shift from labour surplus to labour deficit was clearly the pursuit of industrialization and primarily of export-oriented industrialization. The logic of the argument falls within the classical models of economic development discussed in Chapter 1. During the earlier phase of stagnant real wages, people may choose to migrate overseas to achieve higher real return. As the economy moves into the labour-deficit phase, having passed the Lewisian turning-point, there is a shift from net emigration towards net immigration (Abella 1994a, Fields 1994).

A transition from emigration towards immigration was observed for old core countries in Chapter 3 but, among the new core economies, despite the reality of structural economic change, the transition has been a relatively faint echo; neither the emigration nor the later immigration has been as intense as in Europe. In addition, the impression that there has been a shift in migration in response to changing economic conditions simplifies what has been a much more complicated process. Indeed, there has been a shift in some countries, but political as well as economic factors need to be given full consideration in any explanation. In the case of Japan, late nineteenth- and early twentieth-century migration was both generated and suppressed by policies in receiving countries. Since the 1960s, after Japan passed the turning-point, the number of Japanese residents overseas increased sharply from 241,102 in

1960 to 620,174 in 1990 and 687,579 in 1993 (Watanabe 1994, Japan 1996). These migrants reflect Japan's investment overseas and represent, to a large extent, skilled personnel sent on assignment to oversee Japanese plants and activities around the world, although primarily elsewhere in Asia.

Immigration to Japan has increased. The number of foreign nationals registered as staying more than 90 days in Japan at the end of 1993 was 1.3 million, which was an increase of 40 per cent over the figure of the previous five years. Over half of these were from the Korean peninsula (including North Korea), although the absolute number of Koreans has declined since 1991, as numbers of Chinese, Brazilian, Filipino and Peruvian workers, among others, have increased (Kunieda 1996: 198). In addition to these legal migrants, there are those who have overstayed their visas since they made contact with the underground market. For a lively personal description of what it is like to be a participant in these labour markets in Japan, see Ventura (1992).

South Korea shows perhaps the clearest transition in migration among the new core economies in the patterns of its population movement since the Korean War. Contract labour migration, mainly to the Middle East, developed in the 1970s, reaching a peak during the first half of the 1980s and subsequently falling away in the second half of that decade (Table 2). Settler migration, primarily to the United States, developed from the 1950s, when only 6231 migrated over the whole decade, to reach 333,746 in the decade of the 1980s (INS 1994) but appeared to peak around 1986–7 at over 35,000 for the year, after which there was a marked decline to 15,417 by 1994 (Annexe Table 2). Return migration from the United States is also likely to have increased over this period. Immigration to South Korea, however, rose markedly from the mid-1980s. There are no accurate figures for the number of foreigners working in South Korea but there were probably 100,000 foreign workers, mostly Chinese-Koreans, Chinese, Filipinos, Bangladeshis and Nepalese in the country in mid-1994. About half were estimated to be illegal overstayers and the balance was made up of trainees and skilled professionals (Kang 1995).

Taiwan has shown a transition of sorts, but it is difficult to relate it neatly to labour surpluses and deficits. As mentioned above, it passed the turning-point towards a labour-deficit economy around 1970; yet, from about that period emigration also started to accelerate to reach a peak in the early 1980s. Considerable migration continued throughout that decade, with fluctuations. The trend in the 1990s, some thirty years after the turning-point, seems to be down, there being fewer migrants to the United States than for any year in more than a decade (Annexe Table 2). The migration to Canada from Taiwan, though, still appears to be

Table 2.   Annual outflow of contract migrant workers by country of origin in Asia, 1969–1992 (thousands)

Country of origin	1969	1970	1971	1972	1973	1974	1975	1976	1977	1978	1979	1980	1981	1982	1983	1984	1985	1986	1987	1988	1989	1990	1991	1992
South Asia																								
Bangladesh	–	–	–	–	–	–	–	6.1	15.7	22.8	24.5	30.6	55.8	62.8	59.2	56.8	77.7	68.7	74.0	68.1	101.7	103.8	147.1	188.1
India	–	–	–	–	–	–	–	4.2	22.9	69.0	171.8	236.2	276.0	239.5	225.0	206.0	163.0	113.6	125.4	169.9	126.8	143.6	117.5	416.8
Pakistan	–	–	3.5	4.5	12.3	16.3	23.1	41.7	140.4	130.5	125.5	129.8	168.4	142.9	128.2	100.4	88.5	62.6	69.6	84.8	98.7	115.5	147.3	196.1
Sri Lanka	–	–	–	–	–	–	–	0.5	5.6	8.1	9.4	7.6	14.2	22.4	17.8	15.7	12.4	15.8	15.5	18.4	24.7	42.7	65.0	–
Southeast and East Asia																								
Indonesia	–	–	–	–	–	–	–	1.9	2.9	8.2	10.4	16.2	17.9	21.1	29.0	37.9	56.7	65.5	59.4	64.0	84.1	86.3	149.8	129.8
South Korea	–	–	–	–	–	14.5	21.0	47.7	69.6	102.0	121.0	146.4	175.1	196.9	187.8	152.7	120.2	95.3	86.3	83.0	63.6	55.8	45.7	34.6
Philippines	3.7	1.9	1.9	14.4	26.4	32.7	36.0	47.8	70.4	88.2	137.3	214.6	266.2	314.3	434.2	425.1	389.2	414.5	496.9	477.8	523.0	598.8	701.8	549.7
Thailand	–	–	–	–	0.3	–	1.0	1.3	3.9	14.7	10.6	21.5	26.7	108.5	68.5	75.0	69.7	85.7	85.5	118.6	125.3	63.2	63.8	81.7

Source: International Labour Organization, *Statistical report 1990*, Bangkok, Regional Office for Asia and the Pacific, supplemented by data sheets submitted to the ILO by countries.

growing. The major trend has been a marked increase in immigration to meet the labour deficit. By the end of 1995, there were almost 200,000 legal workers in Taiwan, fully two thirds of whom were from Thailand, with the majority in manufacturing and, like South Korea, in small-scale enterprises (Tsay 1995). In addition, there are also likely to be at least 20,000 illegal workers in Taiwan and the total number of foreign workers has perhaps quadrupled over the past five years.

It is also difficult to see a neat transition in migration for Singapore as the small island republic has always depended heavily upon external sources of labour. In the early 1970s, there were over 100,000 foreign-worker permit-holders in Singapore, accounting for 12.5 per cent of the labour force (Pang and Lim 1982). There has been a continual inflow of short-term labour since then, which has fluctuated depending upon economic conditions and the nature of the demand. In the early 1980s, the total foreign workforce reached 'well over 200,000 or more than 15 per cent of the employed workforce' (Hui 1992) but numbers declined with the global recession and as Singapore shifted its demand from construction to manufacturing. Non-residents in 1990, who would comprise both skilled and unskilled migrants and their families, made up only about 10 per cent of the total population, and a somewhat higher proportion of the labour force, probably around 11 to 12 per cent (Pang 1991, Chiew 1995). Then, the construction boom of the 1990s saw a resurgence in the demand for foreign labour. Recent estimates suggest that the number of foreign workers increased to around 350,000 in 1995, representing 20 per cent of the labour force. Some 95.5 per cent of the general labourers in construction were foreign workers, the majority from 'non-traditional' (that is, not from Malaysian) sources, and primarily from Thailand and Bangladesh (Wong 1996).

An emigration of Singaporeans, however, has taken place throughout this recent period of labour shortages and, although accurate data are difficult to obtain, it may have peaked during the late 1980s, when about 40,000 left between 1986 and 1990 (Cheung cited in Hui 1992). It is a minority who want to leave, but these are the better educated, and they appear to wish to go for a variety of reasons that may not be primarily economic (see Yap 1991, Tan and Chiew 1995). Thus, again, immigration and emigration are linked, rather than one replacing the other in a general structured shift. The Hong Kong case provides additional insight into recent changes in the labour market and patterns of migration.

A migration transition: the Hong Kong case

The above arguments emphasize the structural economic change (economic development) that has unquestionably played a major role in shaping the changing patterns of migration. Yet, emigration and immigration are not necessarily simple responses to the macrolevel measures of labour surplus and labour deficit. As I have shown for Hong Kong (Skeldon 1994), although there was emigration during the period of labour surplus in the 1950s, this did not come from the densely populated urban areas where there was considerable under- and unemployment at the time. Rather, it came from the villages of the New Territories, where structural transformation was taking place because of the decline of the rice economy and where the traditional way of life was being changed. More important was the fact that, at the time, few places in the world would accept ethnic Chinese as migrants. One of those countries was Britain, which allowed access to all Commonwealth citizens. The villagers of the New Territories were among the few in Hong Kong who could conclusively demonstrate that they had been born on British-administered territory and thus qualified for full British passports: a large proportion, probably the vast majority, in the urban areas had been born in China and thus did not qualify.

The New Territories villagers were also able to take advantage of two other fortuitous factors which gave rise to a wave of migration to Britain from the late 1950s through the following decade. The initial linkages were established by Hong Kong men through their recruitment as cooks on Royal Navy and British merchant marine ships. Some were paid off or jumped ship in British ports. The second factor was the demand in Britain for 'exotic' foods in the post-war prosperity of the late 1950s and 1960s, which the Chinese restaurant or 'takeaway' was able to satisfy. The chain migration of poorly educated rice farmers to establish restaurants all over Britain and the virtual depopulation of certain villages is a fascinating vignette of migration and development that has been well told elsewhere (Watson 1975, Baker 1994). That phase was effectively ended once Britain progressively tightened its immigration laws after 1962 to slow the migration from Commonwealth countries.

Hong Kong passed the turning-point from a labour-surplus to a labour-deficit economy some time during the 1960s. The emigration from the New Territories slowed but more as a result of Britain's immigration policy (Chapter 3) than of increasing local demand for labour. There was indeed a significant wave of immigration to Hong Kong following the economic liberalization of the Deng Xiaoping era in China, when net migration accounted for 58 per cent of Hong Kong's total population growth between 1976 and 1981. While these influxes strained Hong Kong's ability to provide adequate housing and services, the addition to the labour force was welcomed by the colony's industrialists; it was 'whispered' that the massive immigration was not unrelated to Hong Kong's growing need for labour (Turner 1980: 55). Yet, during this decade, emigration from the urban parts of Hong Kong also began.

The changed laws in Canada, the United States and Australia from the 1960s, discussed in Chapter 3, facilitated this new development. There was a surge of movement, particularly to Canada, in the early 1970s but, generally, levels remained fairly low because of rising real wages in Hong Kong. Nevertheless, emigration intensified after 1986, precisely at the time that the labour shortage was beginning to be most acutely felt. Numbers leaving increased from around 20,000 a year to well over 60,000 a year within four years and have remained at that level. The easy answer to this apparent anomaly is that it was due to the unique political factors associated with the return of Hong Kong to Chinese sovereignty in mid-1997. However, as I have shown elsewhere, the political factor is but a partial explanation (Skeldon 1990–91). Part of the explanation has to be seen in the demand for skilled labour in the ageing old core countries that was discussed in Chapter 3. It was also no accident that emigration from Hong Kong began to surge just after Australia and Canada began increasing their overall annual intakes following the recession of the first half of the 1980s.

Another part of the explanation lies in the fact that the people leaving are doing so in order to expand business or professional opportunities overseas. The migration is, indeed, a risk-minimization strategy, but that risk may not be mainly political; to establish enterprises with access to markets at different stages in a business cycle makes sound commercial sense. As seen in Chapter 3, the Hong Kong and Taiwan migrants have dominated the business migration programmes of Australia, Canada and New Zealand. They are part of the outflow of rapidly developing countries who wish to expand their activities more widely. The numbers of migrants involved in these programmes may not be particularly large. Much more important is their wealth and dynamism. The overseas Chinese, in particular, have proved to be an especially vibrant group, and they have played a critical role in the development of Southeast Asia, where they have dominated investment and effectively powered economic growth in the region (Wu and Wu 1980, Redding 1990). Their recent migration around the Pacific rim extends these business networks and projects the Chinese overseas into a formidable global force that will have not only economic but also political ramifications in both new and old core areas (Skeldon 1997c). For example, in the United States alone, the number of businesses owned by ethnic Chinese almost tripled between 1978 and 1987, with annual receipts reaching $US9.6 billion in 1987 (Gall and Gall 1993).

Immigration into Hong Kong also accelerated virtually at the same time as the emigration. Legal migrants from China, highly skilled expatriates from other parts of Asia and from the old core tier, as well as domestic workers and labourers all came in increasing numbers. By 1993, more than 100,000 a year were entering the British colony and the stock of foreign nationals had increased from around 170,000 in 1986–7 to over 415,000 at the end of 1995. The labour deficit clearly explains the increasing immigration but, to account for the total pattern of migration, we need to consider broader economic issues and to incorporate political factors. Hong Kong's labour importation scheme, for example, which allowed up to 25,000 workers into the colony, a tiny proportion of the total labour force, had to be scaled back to 5000 in 1996 as a result of pressure from domestic workers' organizations. Thus, although there has been a structural shift in the economies of the new (and old) core countries over the last twenty to thirty years, these cannot simply be associated with a trend from net emigration to net immigration revolving round a single turning-point. There have been a series of turning-points differentiating various types of migrants, not just inmigrants and outmigrants, and the relative balance between emigration and immigration shifts according to global and local political and economic factors.

Consequences of labour shortages

There have been marked increases in immigration to all the new core economies over recent years but, with the exception of Singapore, these migrants represent a very small proportion, generally less than 2 or 3 per cent of the labour force and an even smaller proportion of the total population. From the perspective of the old core tier, public concerns about immigration in the new core and the need for immigration control appear somewhat overstated.

Unlike the old core tier, Japan and South Korea have resorted neither to allowing foreigners in to settle nor to relying on large numbers of guest workers. They have, in the words of Cornelius (1994), 'the illusion of immigration control' whereby authorities aver that they are maintaining closed-door policies to preserve the ethnic homogeneity of their nations, while at the same time allowing increasing numbers to enter the country under various guises. In Japan, as in South Korea, increasing numbers enter as trainees to be absorbed into more or less full-time employment. Unlike the old core tier, the number of illegal entrants is relatively small owing to the physical difficulties of going to Japan and South Korea, and to effective border control, but greater numbers of people enter legally and simply overstay their visas, to which the authorities usually turn a blind eye. The potential for mass immigration no doubt exists, particularly if China begins to participate much more in intraregional flows but, as of the mid-1990s and with the exception of tiny Singapore, no switch towards mass immigration is taking place yet, though there are indications that this is a possible future scenario given the rapid ageing of all the new core societies.

Emigration, too, certainly has not ceased, even if it has slowed to certain major destinations such as the United States from South Korea and Taiwan. The much smaller settler flows to Canada, Australia and New Zealand have generally increased. So, too, have the short-term outmovements of Japanese, as seen above, but also of Korean, Taiwanese, Hong Kong and Singapore professionals as their economies have shifted their manufacturing overseas as a response to the labour shortages and increasing wages at home. All the new core areas, with the exception of Singapore, saw an absolute decline in the number of workers engaged in manufacturing from about 1990 as they moved towards service economies (Table 3). There is a fear that these areas are losing their industrial base and 'hollowing out', but manufacturing has become more capital-intensive and the services themselves are necessary to support the new manufacturing rather than developing at the expense of industry (Selya 1994). However, like the old core tier, there are clear trends towards deindustrialization and a more important service economy. These emergent trends can only reinforce the growing importance of the interchange of skilled and professional migrants.

Another response to the increasing labour shortages in all new core economies has been to incorporate women into the labour force, which has had a marked impact on the status of women, especially in Hong Kong and Singapore. A consequence of this incorporation, and of the emergence of an affluent middle-income group, has been a demand for female domestic servants. In Hong Kong, by the end of 1994, there were well over 150,000, mostly from the Philippines but also from Thailand and Sri Lanka, with perhaps at least one quarter of this number in Singapore. Some, indeed, are channelled into illegal activities in other labour markets (Skeldon 1995a), but the majority represent a low-paid but usually well-educated labour force in that they speak English and usually

Table 3. Numbers employed in manufacturing in the so-called 'newly industrialized economies' in East Asia, 1985–1994 (thousands)

	Hong Kong	Singapore	South Korea	Taiwan
1985	918.8	314.2	3,504	2,488
1986	919.2	306.6	3,826	2,614
1987	916.0	338.7	4,416	2,810
1988	870.9	379.1	4,667	2,798
1989	808.9	403.7	4,882	2,803
1990	751.0	447.4	4,911	2,647
1991	715.7	429.6	4,994	2,599
1992	650.6	434.1	4,828	2,587
1993	596.9	429.5	4,652	2,483
1994	570.2	422.5	4,695	2,485

Sources: Hong Kong: *Hong Kong annual digest of statistics*, Hong Kong, Census and Statistics Department, 1995 edition and supplementary information.

Singapore: *Yearbook of statistics Singapore 1994*, Singapore, Department of Statistics.

Taiwan: *Taiwan statistical data book*, Republic of China, Council for Economic Planning and Development.

Korea: *Korea statistical yearbooks*, Republic of Korea, National Statistical Office.

have at least secondary education. The vast majority in Hong Kong (80 per cent) come from Metro Manila or very close to that metropolitan area, with few from isolated rural areas. Although remittances are important for their communities of origin (see Chapter 6), the deskilling of this group of migrants represents a cost both for their country of origin and for the migrants themselves. The exploitation of female labour is pursued in more detail in Chapter 6, but it must be emphasized here that domestic servants can make up a significant proportion of the foreign workers in labour markets in destination areas, over 80 per cent in the case of Hong Kong. Unlike other workers, they may not be subject to quota and are spread throughout the middle-income areas in the new core areas of Singapore and Hong Kong. Valuable information on domestic workers in Hong Kong is contained in AMWC (1991) and Vasquez et al (1995).

The role of students in migration and development

The changing global migration system since 1965 was described in Chapter 2 and the increasing participation of Asian and Latin

American areas of origin was emphasized in the flows. A significant part of the early flows from any origin, and one that appears to persist once others become more numerous, is made up of students. Here again, relatively small numbers of migrants can have profound importance for the development of origin societies in particular, but also of destinations. Even when the exclusion acts were still in effect in North American and Australasian old core areas, small numbers of students were allowed to study there. Returning students came with new ideas, and many of the leaders of the anti-colonial struggle who guided their countries to independence had studied overseas.

Perhaps the first major organized programme to educate students overseas followed the Sino-Japanese War, when some 10,000 Chinese went to study in Japan between 1895 and 1905 (Harrell 1992). The Chinese authorities saw this programme as an opportunity to learn from a more advanced society while the Japanese saw it as a vehicle to train people sympathetic to their regional ambitions. As with the later attempt to train Korean students from early in the century (Weiner 1994: 63ff), the end result was more the promotion of nationalism. The overseas Chinese students were a significant catalyst for the 1911 republican revolution and Sun Yat Sen himself was a returned student from the United States who had also been closely involved with the students in Japan. The Japanese government generally supported the establishment of the republic (Howe 1996: 397).

Asian students in the Paris of the 1920s were also a significant force in later independence or radical political movements. Deng Xiaoping, Zhou Enlai and Ho Chi Minh were only among the most famous to rise to prominence from the Paris national associations of that time, as were Pridi, Phibul and Prayoon, the architects of the Thai revolution of 1932 (Wright 1991: 48–9). Post-Second World War London, too, saw the formation of student associations that produced generations of political leaders. Lee Kuan Yew, father of modern Singapore, gave a lecture in London on the theme of the 'Returned Student' in 1950 when he was only twenty-seven years old (Josey 1971: 28, see also Turnbull 1982: 251).

Clearly, the movement of students is a global pattern not merely applicable to the new core tier. However, students from these areas, and from the Asian parts of the actively expanding core to be discussed in Chapter 5, have figured prominently in international student flows. For example, students from Asia accounted for almost 30 per cent of the 34,232 foreign students at the tertiary level in the United States in the mid-1950s. By 1993/4, they accounted for almost 60 per cent of all foreign students and their numbers had risen to 265,690 (IIE 1994). New core areas represented four of the top seven places of origin and accounted for 28 per cent of all foreign students (Table 4). Similarly, Asian countries dominated the student flows to Canada and Australia. Within the new core area, Japan hosts large numbers of foreign students, many from old core areas as well as from Asia.

Ideally, the student migration system is circular. Students are sent overseas in order to gain experience and knowledge so that they may help their country to develop upon their return. Students are granted temporary entry visas to allow them to stay in the host country for the purpose of study; they are not workers or settlers. However, there is considerable leakage as students decide to stay and become workers or settlers. Host countries themselves may use the student category to allow immigrants to enter the labour market, as we saw in the case of Japan. In the student migrations from Taiwan in the 1950s and 1960s, very few returned, probably 5 to 10 per cent, but by the mid- to late 1980s, by which time the

Table 4.   Foreign students in higher education in the United States, Canada and Australia: totals by major region of origin and selected Asian areas of origin, selected years

	United States		Canada	Australia
	1954–55	1993–94	1992	1994
Total	34,232	449,749	37,478	49,240
Asia	10,175	264,693	17,703	41,654
China		44,381	3,359	3,864
Taiwan	2,553[a]	37,581	392	1,357
Hong Kong		13,752	6,589	8,927
Singapore	–[b]	4,823	1,161	7,116
Japan	1,572	43,770	772	1,290
South Korea	1,197	31,076	249	1,716
Europe	5,205	62,442	5,876	1,188
Latin America (includes Caribbean)	8,446	45,246	3,008	87
Africa	1,234	20,569	6,248	870
Middle East	4,079	29,509	800	747
Others	5,093	27,290	3,843	4,694

Sources: United States: *Open doors: report on international educational exchange, 1993–1994*, New York, Institute of International Education, 1994.

Canada: *International student participation in Canadian education*, Ottawa, Statistics Canada, 1992.

Australia: *Selected higher education student statistics 1994*, Higher Education Division, Department of Employment, Education and Training.

[a] Refers to Hong Kong, Taiwan and Macau.
[b] Not a major source.

island had become an economic powerhouse, return rates had increased to 20 to 25 per cent (Tsai 1988, Hsieh et al 1989). The loss of these talented young people, together with the emigration of other highly educated movers such as we have seen from Hong Kong and Singapore, is often seen as a 'brain drain' and detrimental in some way to an area's development. Yet, these migrations commenced and grew at the same time as these areas embarked upon their periods of most sustained economic growth. As in the case of the old core areas of western Europe, development and outmigration occurred simultaneously. It is difficult to imagine that the rates of growth among the new core tier would have been higher if the migrants had stayed at home.

The exodus of the highly educated has to be balanced against the gain in skills of those who do return and against the benefits that accrue through the linkage between origin and destination areas. In the Taiwan of the early 1990s, not only was the president Lee Tenghui a returned student, but fifteen of his cabinet of twenty-six had also obtained doctoral degrees from western universities (Lin 1994). Similar situations exist at senior levels of government in most of the new core administrations. The types of open policies pursued and the recent trends to liberalize the political systems and make them more democratic are surely not entirely independent of the attitudes and knowledge gained by the returned overseas students. If these students do indeed play a critical role in the formulation of development policy, then an important question, not only in Asia but also globally, refers to China. Though it is now among the most important sources of students for overseas training, official sources estimate that only a small proportion have returned to China. If the future experience of China mirrors that of Taiwan, and increasing numbers go back to play an important role in their society, then one can perhaps envisage China rapidly developing along a western path and becoming an ally of the west 'due to the compatible values held by the elites of the . . . societies' (Lin 1994: 14).

For the old core countries, the education of overseas students has become a multimillion-dollar service industry. Not only do these countries benefit from some of the leading research minds to develop their own advanced programmes but the presence of the students, in some instances, virtually maintains the viability of certain educational institutions. Almost three-quarters of the students at tertiary institutions in the United States in the early 1990s received their funding from sources outside the United States. They contributed $US6.3 billion to the economy in 1993 (IIE 1994: 104). The multiplier effects are significant, with one study estimating that the indirect impact of 4362 international students created 2386 new jobs, $US34.4 million in household income in the local community and $US5.2 million in state tax revenue (cited in IIE 1994: 104). There are thus 'brain gains' with important development implications for both origin and destination areas.

There can, however, be more negative developmental implications of the loss of students and the brain drain. At the other end of the development spectrum, the sub-Saharan African nations participate least in the international student migration system. Only 14,369 students from East, West and central African countries were studying in the United States in 1993–4, just over 3 per cent of the total (IIE 1994). Yet, even if the numbers are small, their loss, if they remain away, can have a significant impact on origin societies, an issue that will be pursued in more detail in Chapter 7. The sudden loss of hundreds of thousands of scientists from the public sector in Russia and of 90,000 Jewish scientists in particular, the elite of the scientific community, may well have constituted 'a factor in the economic decline of the CIS countries' (Rhode 1993: 239). In this case, the exodus was rapid and massive from a stagnant economy, which had very different consequences from the more gradual outmovement and increasing return of trained personnel from the East Asian economies.

Students can also be interpreted more broadly. Those devout and fanatical Muslims who flocked to Afghanistan from around the Arab world in the 1979–89 war against the communist occupation by the Soviet Union were trained in a variety of combat and covert techniques. With the defeat of the Soviet military, the returned 'Afghanis', or those foreigners who fought in the Afghan war, are a formidable, if diffuse, force for terrorist activities around the world, but particularly for the destabilization of regimes in the resource niche and the labour frontier. These Afghan-trained 'alumni' are estimated to include 5000 Saudis, 3000 Yemenis, 2800 Algerians, 2000 Egyptians and perhaps 2000 Palestinians, Jordanians, Lebanese and others (comment by J K Cooley in the *International Herald Tribune*, 30 June 1996).

Returnees, astronauts and bilocal families

The return of students is but one part of circular flows that can have a profound impact in the new core tier. Perhaps the biggest unknown about the new migration system refers to information about the number of returnees. As in the old core, students, settlers and skilled migrants make up the vast majority of those leaving the new core. Students and the skilled are essentially participating in circular systems, even if there may be considerable 'leakage' towards more permanent movements. Settler migrants should be more long-term, even permanent, yet we know that much return movement occurs to the origin societies. It is necessary to emphasize return movement rather than return migration as there is every indication that further movement will occur back to the first destination upon their return. What appears to be evolving is an extensive network of transnational circuits of short-term and more long-term mobility, with participants having two, and possibly more, residences around the Pacific. This type of movement appears to be characteristic of Hong Kong and Taiwan, in particular, but is very likely to be found among the Chinese populations of other parts of Southeast and East Asia as well.

The form that this migration system takes involves the movement of families as settlers to a city in North America or Australasia and then the return of, most commonly, the principal male breadwinner to Hong Kong or Taiwan to continue his employment. Or the return can involve both husband and wife, with the children being left with relatives or even under the supervision of an older child. The first type of movement has become known as the 'astronaut' phenomenon, which felicitously combines the Cantonese expression for 'empty wife' with 'spaceman', associating loneliness and the time spent on aeroplanes commuting across the Pacific at regular intervals. The second type has become known as the 'parachute kids' phenomenon to convey the idea that the children are just 'dropped' at the destinations.

It is extremely difficult to estimate how many people participate in these systems. It is easy to leave the United States or Canada without a record of departure. As yet, there has been no systematic analysis of exit cards from Australia and New Zealand. In addition, the migrants may have more than one travel document, leaving the origin on one and entering the destination on another, making any exact accounting virtually impossible. Clearly, the expense of regular commuting over long distances and the cost of maintaining two homes will restrict this type of behaviour to relatively wealthy groups. Yet, the available evidence suggests that large numbers of people are participating in these transnational circulatory flows. Indirect estimates indicate that at least 30 per cent of Hong Kong migrants to Australia were returning in the early 1990s after establishing initial residence (Kee and Skeldon 1994). The most compelling evidence for the importance of the astronaut phenomenon comes from the sex ratios of the Hong Kong-born at the major destinations. The 30–39 year cohort in Australia, and for Auckland in New Zealand, and the 25–44 year cohort for Vancouver are heavily biased towards women, showing the importance of female-headed households left at the destination. In Australia as a whole, there were only 84 men for every 100 women among the Hong Kong-born aged 30–39 years in 1991.

The comparable figure for Auckland in 1991 was only 70 men for 100 women. In Vancouver, also in 1991, there were 85 men for every 100 women among the Hong Kong-born aged 25–44 years compared with 99 men for every 100 amongst the non-Chinese immigrants. See the chapters in Skeldon (1994) and also Skeldon (1997b). There also appear to be higher proportions of male children among the Hong Kong-born than among native Australians and New Zealanders, suggesting the existence of 'parachute kids' in these places. A study by the University of California at Los Angeles estimated that there were some 40,000 'parachute kids' from Taiwan in the United States in the early 1990s (cited in the *International Herald Tribune*, 25 June 1993).

This type of population movement represents a 'reverse' sojourner pattern, whereby the family is left at the destination rather than the origin, and the breadwinner returns home, perhaps temporarily, to continue the business or profession. The families thus participate in the continuing economic boom in Asia but, at the same time, have a foothold in a politically stable society. Moreover, if growth rates should slow in East Asia, and the theory of business cycles suggests that they will, the migrants will have access to an economy that is likely to be in a different phase of its cycle and also, in the case of North America, the one that makes up the largest market in the world. Some of the returnees may be representatives of Australian or Canadian firms promoting business in the growth areas of China. Foreign companies are likely to be more successful in those areas if they employ staff who understand the culture and who are ethnically at ease in the often byzantine negotiations that are a critical part of the Asian way of business. Increased trade may therefore be a consequence of such return movements. There are, however, obvious costs in these new transpacific extended families, with lonely, bored housewives dealing with children who are growing up in different cultures and environments (Lam 1994). The husbands, too, may establish new liaisons upon their return to Hong Kong or Taiwan, leading to divorce and further alienation between members of the original family.

Although this pattern of transnational circulation may appear to be unique to parts of East and Southeast Asia, it is only partially so. It is unique in that the family is left at the destination rather than at the origin. The spatial dimension of dual-career households has only recently become an issue in the most developed countries (Green 1995). One aspect is the growing number of married professionals who, at certain stages of their life cycle, pursue their careers in two different locations, coming together weekly or less regularly. There need not be two separate careers involved either. The wish to keep children in particular schools or the sheer impossibility of selling one home and buying another in a stagnant housing market also creates pressures in the old core tier that lead to spatially extended families. Nor is this process necessarily recent. The English public (that is, private) school was an institution that could deal with 'parachute kids' while parents were serving in some far-flung part of empire, and army or navy wives will be familiar with many of the pressures experienced by astronaut families. Modern transport and internal and international population mobility are facilitating spatially separate, dual-career households among skilled and professional groups in old and new core tier alike.

Urban extension or counterurbanization?

The history of recent spatial distribution of population in the new core tier has essentially been one of concentration in huge cities. In the case of Japan, 'no other major country has experienced so dramatic and thorough a transformation, such rapid urban and industrial growth, or comparable concentration of population and industry in major metropolitan areas and a highly localized industrial belt' (Harris 1982: 50). Tokyo has emerged as a global city, with many of the characteristics of other such cities in the old core (Sassen 1991). Hong Kong is also potentially in this class, given its recent structural transformation discussed above, and Singapore, Taipei and Seoul are major regional centres. The expansion of some of these urban centres into their hinterlands and the creation of urban corridors of development will be discussed in Chapter 5, but here we need to examine whether there has been, along with their structural transformation, any trend towards an urban deconcentration such as that seen in the old core areas. About half of Japan's urban population is concentrated in just three metropolitan regions, Tokyo, Osaka and Nagoya (Tsuya and Kuroda 1992: 207). After the intense period of concentration during the 1950s and 1960s, there was indeed a slowing of metropolitan growth in the 1970s. Yet, this did not appear to represent a counterurbanization in the European or American sense: much more it reflected growth on the periphery of the existing metropolitan regions. The process in Japan has been termed the 'doughnut' effect, with most of the fastest-growing areas found in the non-metropolitan regions around the main centres in a *de facto* spatial extension of the metropolitan zones. This process has been reinforced by 'J-turn' migration patterns, which describe original rural-to-urban migrants moving back again, perhaps on retirement, not to the rural areas from which they came, but to the regional town, again often close to the main cities. Patterns of decentralization of this type have also been observed around Seoul and Taipei (Hsieh 1985) and the extensions around Hong Kong and Singapore will be discussed in Chapter 5. The physical constraint of available land is likely to be a significant factor in accounting for a lack of real deconcentration of urban growth in Japan. Again, as in the United States and parts of Europe, the 1980s in Japan have seen the resumption of a slow trend towards an increase in urban concentration.

Discussion

The economies of the new core tier have achieved levels of development that have surpassed those of several of the old core countries. Their political systems are developing mechanisms for the smooth transfer of power within more open governments, even if popular participation is not as deeply embedded as in the old core. The migration patterns over the last century have shown some parallels with those of the old core but many differences. Even the parallels are variants rather than closely similar. For example, the prevalence of pre-industrial systems of human circulation and the period of intense concentration in large urban centres are the two most obvious parallels. The latter was compressed in time and more rapid than in the old core. There has been a relatively faint echo of urban turnaround and deconcentration, with peripheral metropolitan growth more a characteristic of the new core than the vibrancy of small separate urban centres.

The general trend in transnational moves has been from emigration towards immigration, but neither of these flows has been as intense as in the old core, although in some cases both were numerically very significant. In large part, the more muted patterns were due to the fact that the new core was a relative latecomer to development. The new core areas had to operate within constraints laid down by old core areas, with much of the migration actually controlled by British or American interests in the nineteenth century. With a few exceptions, Asians were entering destinations controlled by others rather than areas where they could establish their own authority. There have been mass migrations between the new core areas but primarily associated with Japanese imperial ambition and its immediate aftermath; the more recent trends towards immigration have been less accentuated than in the old core.

Despite the differences, there does appear to be a trend towards convergence (see also Sassen 1995). Ageing societies and economic restructuring towards service economies within the context of an international division of labour have brought demands for particular types of labour on various points of the skill spectrum but at the same time a wish to regulate the immigration of labour tightly. Migration for settlement is limited in the new core mainly to people from the same ethnic group as the host society – Chinese in Hong Kong, Singapore and Taiwan, and Brazilian or Peruvian Japanese in Japan, for example – and the concept of multiculturalism is virtually unknown, except in Singapore. All other immigrants are viewed as short-term workers who will eventually return home. The immigration, irrespective of its duration, is nevertheless bringing a greater diversity to societies that have perceived themselves to be homogeneous, opening the door, at least marginally, to the polyethnicity that is characterizing the old core. This trend will almost certainly continue in the new core.

Chapter 5

Core extensions and potential cores

The third development tier is the spatial extension of the first two tiers: in some areas it is a direct expansion into contiguous areas so that there is no real dividing line between core and core extension. In other areas, it is an outlier, physically far from core areas but nevertheless intimately tied to them through patterns of investment, trade and population transfers. Some of the areas included in this tier are experiencing rapid economic growth, others have shown stagnant, even negative, growth; some areas are highly urbanized, others are much less so, but are urbanizing rapidly. Often only part of a country is included under this development tier: Brazil, China, India or Mexico certainly cannot be considered as consisting of a single development region, even at the global level. Intra-country differences have largely been ignored up to this point as, even where such differences existed, the strength of the state and national institutions appeared to justify their consideration as a single unit. From here onward, this assumption will often not be possible and there will be a continual tension between the need to look at only one part of the state and the need to consider certain types of information which are available only at the level of the state.

This development tier has been divided into three parts: actively expanding extensions of the core areas; potential core areas; and the restructuring core. The actively expanding core areas are found in east Asia in coastal parts of China; in Southeast Asia in much of Peninsular Malaysia and central Thailand; and in four separate regions in Latin America: southern Brazil, central Argentina, northern Venezuela, and central and northern Mexico. The potential core areas are those areas that appear most likely to develop as cores over the medium term. They are found in western and southern India; in west Asia in Israel, and in southern Africa, mainly in South Africa. Finally, to these must be added the restructuring core, which includes parts of the former Soviet system which, since 1989, have begun to rejoin a global system. Other possible future cores can be identified, but these will be discussed in Chapter 6. There is thus tremendous diversity within a single development category in terms of culture, historical experience and political systems, but it is justifiable to consider them together as these areas are likely to include those that may join the more prosperous core areas over the medium term. This does not mean that all will do so; nor does it mean that there are other areas outside the tier that will not achieve prosperity (see Chapter 7), but when we come to assess global development patterns in the middle of the twenty-first century, it is likely that these areas will be included in any new core tier.

Within this development tier are some of the largest cities in the world: São Paulo, Mexico City, Buenos Aires, Beijing, Shanghai, Bombay and Bangkok, with all their attendant environmental and infrastructural problems. It is in this development tier that urban expansion into enormous agglomerations or urban corridors is most marked. City regions are nevertheless dependent upon capital and highly skilled labour from the tier above and upon more low-skilled labour from the tier below.

Quite apart from lower levels of economic development in this tier compared with the core areas, the political institutions are various, from functioning democracy of a type such as India to authoritarian states such as China. All, however, are subject to internal pressures of secession and of succession, or both, and long-term stability is much less assured than in the core areas. From the point of view of global security, these areas, or at least the states in which they are found, will play a critical role in determining just which directions future development will take. Politically and economically, this tier is on the frontier of the free-market core regions, not necessarily physically, but through the nature of the linkages with them of which migration is an important part.

This image of frontier becomes more vivid upon the demise of the socialist system as a viable economic alternative. The critical issues revolve around how far the free market has penetrated beyond the old and new core tiers. That penetration is extremely patchy, and large parts of the world, while certainly not excluded from the global economy, are participating much more weakly in it. For example, if we look at foreign direct investment as a key to the intensity of the linkage to a global system, we find that some 78 per cent of the stock of foreign direct investment in 1994 was concentrated in the two core tiers (United Nations 1995c). A further 15 per cent was to be found in the countries that cover the extending core tier, with only 7 per cent in the rest of the world. Despite the momentous changes that have occurred over the past fifteen years, the distribution of foreign stock invested has not changed significantly, even though its magnitude has increased almost fivefold since 1980. In that year, 76 per cent of the stock invested was in the old and new cores, with 15 per cent in the extending core. Thus, consolidation of an established pattern rather than extension might be a more accurate description. Claims of the expansion of multinational activity leading to the erosion of the nation state, with the latter dysfunctional in some way for organization of economic activity (Ohmae 1995), have to be treated with some caution as the expansion of such activity is precisely where the state appears to be strongest. Nevertheless, there is perhaps some basis for the claims, and it is in this expanding core development tier, often on the boundary with the cores, that we can see the emergence of development regions, or growth triangles, that stretch across national borders. This regionalization is becoming significant in East and Southeast Asia (Thant, Tang and Kakazu 1994) and it is there that the discussion of migration in this development tier will begin.

The expansion of the new core

Ageing societies and economic restructuring as the new core areas moved towards service economies were described in Chapter 4. One of the responses to rising labour costs at home was to export labour-intensive industries overseas. Surplus capital generated at home could be invested overseas to maintain these industries in profitable environments and to promote new ventures. Japan's investment in Southeast Asia, some $US27 billion between 1951 and 1990, particularly in Indonesia and more recently in Thailand, has been an important driving force in the economic development of those countries. So, too, has Taiwan's investment in China. One estimate places the amount of investment from Taiwan in China at more than $US5 billion between 1987 and 1994 alone (Myers 1995), with much of the investment concentrated in Guangdong Province, although it is also found immediately across the strait in Fujian Province and in Shanghai. However, perhaps the most dramatic examples of expansion have been Hong Kong's movement into southern China, and Singapore's movement into Malaysia. Labour and rental costs just across the border in Guangdong Province are a fraction of those of Hong Kong, even in the Special Economic Zone of Shenzhen adjacent to Hong Kong. The 'industrial colony' (Hopkins 1971) of only a few years ago, Hong Kong has now moved its manufacturing base into China, mainly into the adjacent parts of Guangdong Province. By 1995, there were well over 3 million people in China working in Hong Kong-controlled enterprises, a total many times that of the declining number of people employed in secondary activities in Hong Kong. Hong Kong, and to a lesser extent Taipei, act as the financial and service centres. A functionally integrated megalopolis has been created, stretching around the Pearl River Delta from Hong Kong in the east through Guangzhou to Macau in the west (Skeldon 1997a), which has idealized functional divisions as shown in Fig. 5.1. Its total population exceeds 25 million people and it thus forms one of the largest urban areas in the world, though it is not included as such in published statistics.

Around Singapore, the scale is somewhat smaller, but the impact has nevertheless been significant. In the Singapore growth triangle formed by Singapore, Batam (in the Riau Islands of Indonesia) and Johor State, some 75,000 jobs were created in Johor by Singapore investment alone over the decade of the 1980s, and the population of Batam had increased from a few thousand in the 1970s to well over 100,000 by 1991 (Kumar 1994). The total population of the triangle in the mid-1990s is probably well in excess of 3.5 million.

The implications for migration are twofold. First, there is an inflow of professionals from Hong Kong, Singapore, Taiwan and Japan to oversee factories in the development zones. Precise numbers are very difficult to come by, but there are probably at least 100,000 from Taiwan in China alone and some 64,200 workers based in Hong Kong, mostly highly skilled, had worked in China at some time in the twelve-month period to mid-1992 (Hong Kong 1992). Second, and numerically much more impor-tant, is the inflow of workers from the immediate hinterlands and beyond. If there is one common characteristic among all the diversity of this development tier, it is the concentration of popu-lation in large cities. There is intense rural-to-urban migration and urban-to-urban movement up the hierarchy. However, whether the form of the resulting mega-urban regions is comparable to those found in old core countries has been a subject of debate. Terry McGee (1991) and Norton Ginsburg (1991) have separately argued for a distinct pattern of urbanization in which urban and rural are much more closely integrated than in the historical experience of Europe and North America. In particular, the growth of urban systems within rice economies produced a distinctive blend of urban industrial and rural agricultural landscape that McGee termed *desakota*, coined from the Indonesian terms for town (*kota*) and *desa* (village), which links the main urban nuclei in corridors of development.

There is no question that the ecological context in which the urbanization is occurring in Asia is different from that of Europe. A landscape of city growth among rice paddies is going to be very different from one evolving among the rain-fed temperate agri-culture of northern latitudes. Yet, as we saw in Chapter 3, there was intense circulation between rural and urban areas in the historical experience of Europe and, there too, agriculture was not excluded from urban activities in the early growth of cities (see

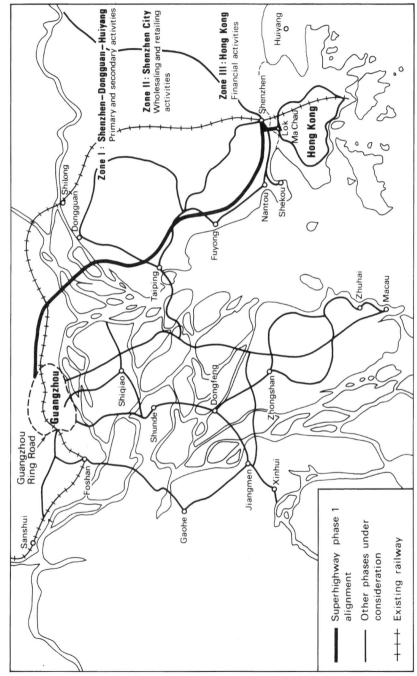

Fig. 5.1 Idealized expansion of the core: the functional urban region around Hong Kong

Zone I : Shenzhen—Dongguan—Huiyang
Primary and secondary activities

Zone II: Shenzhen City
Wholesaling and retailing activities

Zone III: Hong Kong
Financial activities

Huiyang

Shenzhen

Lok Ma Chau

Hong Kong

Shilong

Dongguan

Nantou

Shekou

Fuyong

Taiping

Zhuhai

Macau

Guangzhou Ring Road

Guangzhou

Shiqiao

Dongfeng

Zhongshan

Shunde

Foshan

Sanshui

Gaohe

Jiangmen

Xinhui

Superhighway phase 1 alignment

Other phases under consideration

Existing railway

Source: Nomura Research Institute (Hong Kong)

Langton and Hoppe 1990, for example). As migration towards the urban areas intensifies, we can expect agricultural areas to be transformed and then to disappear as the demand for industrial land gradually increases. The example of the immediate hinterland of Hong Kong is instructive. Rice as a labour-intensive crop virtually disappeared from the rural parts of Hong Kong by the early 1960s in the face of the rising wages offered by the growing industrial sector. More profitable vegetable- and flower-growing replaced the rice in areas of favourable location, but these were grown by immigrant farmers from mainland China who were not only experienced vegetable-farmers but also willing to work for lower wages (Topley 1964, Watson 1975). With the structural transformation towards a service economy described in Chapter 4, the areas of vegetable gardening declined as the standards of education of the children of immigrant vegetable-farmers rose, and as salaries in the urban areas drew ever larger numbers into these activities. More and more of the agriculture was pushed into southern China, but there too the rice economy has been under pressure for some considerable time. Within the Pearl River Delta region alone, the area of arable land declined by 14 per cent between 1985 and 1991 and, in the counties closest to Hong Kong and Guangzhou, the decline was 30 per cent or higher (data cited in Wu 1997). At the national level, the losses of crop land have been dramatic. Between 1957 and 1988, official figures give an average of some 520,000 ha of agricultural land lost every year to rural and urban modernization, and the real figure could be three times this (Smil 1993: 56–7). Thus, the mix of agricultural and industrial land uses observed in the *desakota* pattern may be but a transient phase towards a more intense urban industrial complex not too dissimilar to the old core counterparts. As the Asian economies are opened to more free-market forces, the progressive, and seemingly inexorable, rise in land and labour costs brings about a structural shift towards more high-value activities, with agriculture being pushed farther and farther into the periphery of this development tier.

Development in East and Southeast Asia has proceeded at such a pace that regional labour shortages have emerged in virtually all parts of this expanding core tier which have generated the tremendous movement into both urban and rural sectors. This movement is both internal and international, with its origin primarily in the labour frontier areas to be described in Chapter 6. The volume of this movement can be massive and in no place more so than in China. Between the 1950s and about 1978, the Chinese authorities were, in large part, able to control internal movements of population. This often involved the 'sending down' of millions of people from urban to rural areas. In the campaigns of 1966 and

1967, some 20 million educated young people were moved, often to isolated rural areas in the west of the country, where they were supposed to foster development. These forced movements were by no means atypical, and even larger numbers were reported to have been moved in earlier campaigns in the 1950s and early 1960s (Kirkby 1985: 38). These movements, together with restrictions on migration to urban areas, meant that the Chinese authorities were indeed able to limit urban growth during this period and to produce a very different pattern of urbanization and population redistribution from that of most other parts of the less developed world.

Since the reforms implemented from 1979, population mobility and urbanization in China have taken on patterns which are immediately familiar to those who have examined the phenomena elsewhere (Kojima 1995: 121). The date 1979 is almost certainly too rigid a cut-off, and White (1994) has shown that, in the hinterland of Shanghai, increased urbanward migration had begun considerably before then. Nevertheless, 1979 marks a convenient turning-point in the internal migration in China, which has seen persistent and increasing movement towards the largest coastal cities of Beijing-Tianjin, Shanghai and those in the Pearl River Delta region in particular. Net rural-to-urban migration appears to have been in the region of 6–7 million each year during the 1980s, to which must be added the short-term 'floating population', which had probably reached around 70 million by 1989 (Chan 1994). Although total population movement, as measured from the 1990 census of China, shows that males still dominated migration, several of the largest inter-provincial flows towards the coastal cities were female-dominated, reflecting the demand for a cheap, easily manipulable, labour force in the light labour-intensive industries established by Chinese from Hong Kong and Taiwan in these coastal regions.

In Asia, Peninsular Malaysia has emerged as another target for migration. Rather like China, Malaysia, some twenty-five years ago, was seen to represent a pattern of migration and development distinct from that of the core areas given its 'low rate of urbanization and emphasis on frontier agricultural settlement' (Pryor 1982: 26). Since then, a transformation of both the economy and the migration system has taken place, with massive rural-to-urban migration and acute labour shortages in both urban and rural sectors that have been met, at least partially, through immigration primarily from north and west Sumatra and east Java in Indonesia (Hugo 1993).

By 1991, over half of Malaysia's population was classified as urban, up from 27 per cent in 1971, and by the early 1990s some 18 per cent of the labour force was estimated to be from overseas sources, over 1 million legal foreign workers. Pillai and Yusof (1996) estimated that, during the period 1991–5, employment grew at 3.2 per cent per annum, whereas labour force growth was only 2.7 per cent per annum.

Initially, the internal rural-to-urban migration in Malaysia created a labour vacuum in the rural areas, particularly in the plantation sector, that was filled by males coming in from Indonesia. Later, as the economy took off into double-digit growth from the late 1980s, more and more overseas migrants entered the urban sector directly, mainly in the construction industry. Again, the majority of these came from Indonesia, although there were also workers from Thailand. The demand for labour is not simply at the lower levels and there are estimates that there will be a shortfall of some 34,000 engineers and technicians by the end of the century (Pillai and Yusof 1996).

Thailand, but essentially the Bangkok metropolitan region, has emerged as the most recent node of rapid development in Asia. The metropolitan area accounts for just over half of the total GDP and development has occurred so fast that only about 55 per cent of the demand for engineers, for example, can be met locally. Trained personnel are being brought in from India, the Philippines and Pakistan to fill the vacuum (Yongyuth 1996). Much more important in terms of absolute numbers of migrants are the unskilled labourers who are coming in illegally from Burma, Laos, Cambodia, China and South Asian countries. They are undercutting local wages and filling jobs in rural and urban areas that the increasingly educated Thais are no longer willing to do. The majority of these migrants, perhaps some 525,000 by late 1994, come from Burma and are found mainly in border areas, although they are also to be found in the capital and in fast-growing tourist centres such as Pattaya. Both men, going into agriculture, construction or labour-intensive urban activities, and women, entering as maids or prostitutes, are involved in this illegal migration.

In these three parts of Asia, the migration has been principally a product of the 1980s, increasing towards the end of that decade. Clearly, before this time there was also inmigration – to Thailand and Malaysia from southern China in the late nineteenth and early twentieth centuries (see Chapter 4), for example. Movements from Indonesia to the Malay peninsula date back over 500 years, though temporary labour migration only became important during the colonial period in the late nineteenth century (Hugo 1993). Although the present movements are very much a product of the rapid economic development from about 1986, and they followed a long period of relative immobility across national frontiers from the 1920s, there are strong historical antecedents that are not without their influence on the current patterns.

As in the case of the new core areas discussed in Chapter 4, there has again been no neat transition in mobility from emigration to immigration in the rapidly expanding core. Emigration certainly pre-dated immigration in Thailand but it continues to increase at the same time as immigration. Thais started going to the Middle East as contract labourers from the 1970s, although the numbers did not become significant until the 1980s. The numbers peaked in the late 1980s, at around 78,000 per annum, after which there was a gradual shift in the direction of movement towards regional destinations. As we saw in Chapter 4, Thais are going as construction workers to Singapore and into manufacturing in Taiwan; others are going into agriculture in Malaysia and as 'entertainers' to Japan. Estimates suggest a growth in the number of Thai workers going abroad from fewer than 4000 in 1977, to around 63,000 in 1990, to almost 200,000 in 1995 (Yongyuth 1996).

In China, too, the recent increasing inmigration and emigration have developed simultaneously and the emigration is occurring precisely from the most developed parts of southern and coastal China, the areas to which the inmigrants are primarily moving (Skeldon 1996). The case of Malaysia also showed simultaneous rising emigration and immigration up to the early 1990s and Hugo (1993: 53) cites sources that indicate some 25,000 illegal Malays in Japan and 90,000 in Singapore, with 41,500 having legally settled in Australia during the decade of the 1980s. After 1992, the numbers of overstayers from Malaysia in Japan dropped from 38,500 to 14,500 in 1995 (Iguchi 1996) and the numbers of migrants going to Australia legally have also slowed sharply. In part, the latter will have been due to Australia's own cutback in annual intake but, in part too, it has been due to the continuing economic growth in Malaysia and the economic slowdown in both Australia and Japan. Return movements from Australia to Malaysia also appear to have increased (Hugo 1994a) but whether this is an 'astronaut' pattern similar to that discussed in Chapter 4 is impossible to conclude at this stage. While immigration has grown, emigration has slowed but, to conclude that there has been a 'turnaround' as such is somewhat deceptive as, although accurate data are not available, it is likely that, as in the case of Hong Kong discussed in Chapter 4, there was never any period of net emigration in the recent history of Malaysia.

Thus, while the underlying economic trends cannot be ignored, policies adopted in the areas of origin that are allowing more people to move, as in the case of China, and policies at destination countries that are permitting peoples of different national origin to enter, both legally and illegally (or perhaps the extent to which authorities are prepared to turn a blind eye to illegal migrants might be more exact), must be taken into account in any explanation of the migration. The changing patterns of movement cannot simply be explained as a reaction to shifting labour surplus and labour deficit. Emigration is clearly part of the recent development experience of this tier and occurs from precisely those areas where there may be labour shortages.

Although emigration is accelerating from parts of Asia, the numbers leaving are still very small compared with the base populations. Of course, the impact that migrant groups can have on destination societies and economies is not simply a matter of numbers. Relative education, wealth and entrepreneurial drive can translate into major impacts for relatively small numbers of migrants. Nevertheless, if we are to talk of 'the age of migration' (Castles and Miller 1993), then numbers must be a part of the equation, and particularly in Asia where most of the world's population are to be found. Migration is increasing, but it is unlikely to reach proportions similar to that from Europe 100 years ago. The coming wave of population movement is likely to be contained primarily within national boundaries or within regional catchment areas of transnational movement such as those identified above.

The expansion of the old core

The old core has shown neither the recent dynamic economic growth of the new core, nor perhaps as clear and visible a sign of its spatial extension. Nevertheless, that expansion has occurred, though it is much more widely dispersed throughout the economies on the periphery. An obvious physical parallel is along the United States–Mexican border, where the *maquiladora* industries have been established. These industries were set up specifically to take advantage of Mexico's low-cost and abundant labour and the close proximity to the United States market. American parts could be sent for assembly to a neighbouring area where supervision was relatively simple and transport costs were lower than having them sent to Asia; the finished products were then reimported back into the United States market. As in the case of Guangdong Province or Johor State, overseas companies set up branch plants, just across the border from the core, for export, not for the domestic market in the countries in which the industries were located.

The expansion into Mexico generally pre-dates that of the new core expansion by over a decade, although Japan was experimenting with overseas factories in Southeast Asia at about the same time or even earlier. In 1967, there were only 72 *maquiladora* factories across the border in Mexico, with some 4000 employees. By the mid-1970s, there were 455 factories with 76,000 employees and by 1992 there were over 2000 factories and almost half a million employees. Originally established in the border town of Nuevo Laredo in the mid-1960s, the distribution of factories spread as their numbers increased, not only further along the border and deeper into northern Mexico, but well into the centre of the country as far as Guadalajara, San Luis Potosí, Torreón, and the capital city itself. Perhaps the best overall analysis of the vast literature on these industries is Sklair (1993) and for a discussion of the evolution of an integrated border metropolis, see Herzog (1991).

The creation of this enclave economy has had a major impact on urban growth on both sides of the border, which has been higher than the average in either Mexico or the United States. Yet, unlike the situation in the expanding core in East Asia, where there has been relatively little onward movement of migrants into the core, the *maquiladora* industries merely serve as an intervening opportunity on the well-trodden path of northward movements into the United States. The origins of that migration out of Mexico show interesting parallels with the early movements out of China: both flows were dominated by males engaging in circular labour, sojourner migration, in jobs amongst which the construction of railroads figured prominently. Later, and to this day, agriculture has provided much short-term employment for Mexican workers. Whereas the Chinese were excluded from the United States from the 1880s onwards, the Mexican circulation continued with fluctuations until the depression of the 1930s stopped the migration. Some half a million Mexicans either returned home voluntarily or were repatriated during the 1930s when preference for all jobs was given to United States citizens (Massey et al 1987: 42). The circular labour migration to the United States, again mainly of males, was resumed with the *bracero* programme from 1942 and lasted until the major changes to the Immigration Act of the mid-1960s described in Chapter 3. By that time, the demand in the United States for labour, again mainly in the agricultural sector, was exceeding the supply of the numbers of *braceros*, and the massive illegal or undocumented migration from Mexico was well under way. By the mid-1980s, there were estimates of over 3 million undocumented Mexicans in the United States (Passel and Woodrow 1987) and there had been over 7 million arrests of illegal Mexicans during the 1970s (cited in Hondagneu-Sotelo 1994: 23).

It is unclear just what impact the expansion of *maquiladora* industrialization might have had on the overall flow of northward migration. The majority of those who were employed in the *maquila* industry appear not to have migrated specifically to obtain *maquila* jobs; they moved either to join their families or to be closer to them. The *maquila* industries may have absorbed some potential movement into the United States (the 'buffer' thesis) rather than attracting large numbers of potential employees, not all of whom could be employed, and who then moved on to the United States (the 'magnet' thesis) (Sklair 1993: 165–6). However, there appear to be different patterns for men and women, with women being more likely to fall into the first category and some young men being attracted to the new industries but, failing to find employment, then moving on across the border. What is clear about the migration is that, like the recent employment in so many parts of the expanding core in Asia, the demand has been primarily for female labour. In the mid-1970s, almost 80 per cent of employees in the *maquiladora* industries were female, although that proportion had dropped to around 60 per cent by the early 1990s (Sklair 1993).

The dominance of men in the migration to the United States has pushed women into the local labour market in the home areas to make ends meet during the prolonged absences of husbands, fathers or sons from the household (Hondagneu-Sotelo 1994, Massey et al 1987). Women in Mexico, as in Latin America generally, have figured prominently in the rural-to-urban migration flows for at least fifty years, and much earlier from some areas. Although males generally appear to be among the earliest migrants from any community (Skeldon 1990), women are not merely passive followers. As Hondagneu-Sotelo (1994) has shown in her analysis of emigration from Mexico, the majority of women migrate fairly independently, and not only young adult single women but wives and daughters, who may move despite the express wishes of husbands and fathers. Any analysis of migration needs to look at the role played by women in the origin as well as the destination areas. In some areas, in Africa and the Pacific, for example, women play the dominant role in agricultural production in their home areas. Their absence would have a greater impact than it would in those areas where they play an equal or more subservient role, as in most of Latin America and Asia. Women may therefore be 'freer' to move, despite patriarchal attitudes, in the latter areas than in those where their absence will make a greater impact on the local economy. This issue of gender relations in migration will be further pursued in Chapter 6.

Like the migration in all other areas examined thus far, the impetus for emigration and for rural-to-urban movement came with the structural transformation of origins and destinations. In China, the reforms of the post-1979 period provided the impetus for an upsurge in population movement. The spread of the responsibility system in the rural areas made clear the surplus of population there and pushed many away from a disintegrating commune system. The opening-up of the coastal areas to foreign investors established the industries that drew upon part of that surplus. In Mexico, the reforms of the Porfirio Días period in the late nineteenth and early twentieth centuries smashed the communal system of landholding, creating the hacienda and the situation where hundreds of thousands of peasants had no access to land; the expansion of activities in the western United States provided more highly paid job openings for manual labour at about the same time (Massey et al 1987). Thus, like the contrast between the old and new cores, as one might expect, the periods during which old and new core expansion has occurred have been very different. In the Asian expanding core, massive rural-to-urban migration has either just occurred, or is occurring now, whereas in Mexico, and in the other parts of Latin America to be discussed below, it took place over a much longer period, reaching its greatest intensity some thirty years ago. For example, Mexico was already classified as 42.6 per cent urban in 1950, reaching 50.7 per cent urban in 1960, around 60 per cent in 1970 and 74 per cent in 1992. No Asian countries outside the core discussed in Chapter 4 even closely approach these later levels (Annexe Table 1). Nevertheless, the expansion of capital in what Sklair (1993) has termed the 'reformation of capitalism' to establish export-oriented industrialization has occurred in both areas within a relatively short period of time. In Mexico, the establishment of this industrialization was within the context of an existing well-developed urban system and a long-established tradition of migration, whereas in Asia the new industrialization and the creation of an urban system through rural-to-urban migration are occurring simultaneously.

The expanding Mexican core extends southwards to include the cities of the central plateau and valleys, including the capital city. Three other expanding core areas can also be identified in Latin America: in northern Venezuela along the Caracas-Valencia axis; in southern Brazil in the Rio de Janeiro–São Paulo axis; and finally around Buenos Aires. All these regions attracted large numbers of migrants from Europe during the period of the Great Migration across the Atlantic discussed in Chapter 3, and movements from Europe continued, with fluctuations, into the post-Second World War period. All these regions experienced massive internal movements from rural areas in the post-1950 period, reaching levels of urbanization equivalent to, or higher than, those of the majority of old and new core areas by the 1990s. Even in 1950, some 62.5 per cent of Argentina's 20 million people lived in urban areas. By 1992, the total population had risen to over 33 million, 87 per cent of whom lived in urban areas and fully one third of that proportion in Buenos Aires. Brazil and Venezuela were 36.2 and 53.8 per cent respectively urban in 1950 and had reached 77 and 91 per cent urban respectively by the early 1990s (all data for 1950 are from Elizaga 1970 and for 1990 from World Bank 1995). Compared with the expanding core in Asia, these Latin American areas are essentially urban societies. Their manufacturing and commercial activity, too, is highly concentrated, though there is a trend towards decentralization within the general zone around the largest cities. Well over half of Brazil's manufacturing is located in São Paulo, some two thirds of Argentina's industrial employment is located in and around Buenos Aires and fully three-quarters of Venezuela's is found along the Valencia-Caracas axis (Gwynne 1987: 124).

Argentina and Venezuela have persisted as major destinations of migration but, recently, from other Latin American countries rather than from Europe. As North America saw a switch in migration from European to Latin American and Asian origins, so the principal South American destinations saw a switch in the origins of their immigration from Europe to, in the case of Venezuela, Colombia and Brazil; and, in the case of Argentina, to Chile, Uruguay, Paraguay, Bolivia and Peru. Not all of this immigration is directed at the expanding core parts of these countries, however, with thousands of Brazilians entering the Amazonian parts of Venezuela illegally to search for gold (Stalker 1994: 227). However, as in Asia, regional migration systems are emerging as the dominant type of movement. Global processes are producing regional systems.

The starkest contrast between Asia and the Latin American expanding core tiers is in the pattern of their recent economic growth. While East and Southeast Asia have seen almost continuous rapid growth over the last quarter of a century, the Latin American pattern has been characterized by marked fluctuations with a prolonged period of stagnation, even negative growth in some sectors during the 1980s, as the Latin American nations have struggled with their accumulated debt. When combined with periods of repressive governments, this stagnation was met in Argentina and Brazil, in particular, by return migration of settlers of European origin, though Argentina and Venezuela appear to have maintained a net immigration throughout this difficult period. Venezuela, unique among the areas under consideration here, still depends heavily upon its oil resources but, despite this apparent advantage, its GNP per capita declined by an average of 1 per cent per annum between 1965 and 1990, with the most acute falls in the early 1980s (Hackett and Summerscale 1995: 637).

Migration in South America thus appears to be but a pale reflection of the dynamism of the East and Southeast Asian systems and there might seem to be little justification for including these three areas in the 'rapidly expanding core' category. Nevertheless, since the mid-1980s, there has been a major change throughout Latin America. Democracies were restored in Argentina and Brazil in 1983 and 1985 respectively, and liberal economic reforms, which opened up the national economies to overseas competition and investment, appear to have led to an upturn in production and trade in the early 1990s. Both Argentina and Venezuela are now concerned about skilled labour shortages and have programmes to attract migrants from eastern Europe (Stalker 1994). Argentina, the seventh richest country in the world in the 1930s, and southern Brazil and northern Venezuela, which both have substantial domestic markets, should be able to drive economic growth in their respective regions, attracting migrants from the surrounding regions into the largest urban industrial complexes. Present attempts to create wider regional units, the South American Common Market (Mercosur) covering Argentina, Brazil, Paraguay and Uruguay, a population of almost 200 million accounting for 60 per cent of Latin America's total output, and the Andean Pact covering Venezuela, Bolivia, Colombia, Ecuador and Peru, may favour the three expanding core regions. However, the attempted political coup in Venezuela in early 1992 and continuing economic problems in that country attest to the fragility of these societies and economies. This absence of political, and ultimately economic, robustness is indeed a common characteristic of this entire development tier. Reliable predictions are notoriously difficult to make, but major political and economic problems are likely to lie ahead in China, in multi-ethnic Malaysia and even in Thailand, where the linchpin of stability is a revered but ageing monarch. Growing disparities in wealth, corruption and a fissile political and civic culture may threaten the continued stability of these areas, leading to a reversal of gains already made. As we will see in more detail in Chapter 7, development and the direction of migration can be reversed.

We have seen the expansion of core activities to the west and south of the new core and to the south of the old core in North America, but it is difficult to pinpoint expansion from the old core in Europe. Of course, it can be argued that the areas in southern Europe in Spain, Portugal, Italy and Greece represent the earlier expansion of activities from further north, and their transition from labour exporters to labour importers was described in Chapter 3. The expansion of the old core in Europe can perhaps be seen more as a consolidation towards the southern periphery of its own borders. Some attempts have been made to set up economic zones in central and eastern Europe but these tend to have only local impact (United Nations 1991). The general trend is towards net emigration throughout eastern Europe, which is quite different from the centres of attraction that characterize the other areas of this expanding core development tier. The eastern periphery of Europe, including Turkey, is thus best covered in Chapter 6, the labour frontier. Three other areas will be considered in the present chapter. These are much more tentative regional cores and can be termed 'potential core areas'.

Potential core areas

Three areas appear to have the potential to generate rapid growth and draw in population from wider areas, although the critical question whether their potential will ever be realized relates more to politics than to simple economics. The areas are in southern Africa centred on South Africa; in the Middle East centred on Israel; and a discontinuous crescent stretching from northwestern to southeastern India. While it may be invidious to make such a sweeping generalization, it can be said with some justification that South Africa and Israel are, more than any other areas in the world, the creation of migration, though in very different ways.

South Africa was, like other temperate regions around the world, a frontier of European settlement. From the mid-seventeenth century, Dutch, Germans, French, Huguenots and then British settlers established themselves and moved inland in search of land to settle or souls to convert. The northward migration of the Afrikaner farmers from the imposition of British administration, and their battles with Ndebele and Zulu peoples, gave 'The Great Trek', as this migration was known, 'its epic dimensions as the foundation myth of Afrikaner nationalism' (Omer-Cooper 1995: 841). It is worth recalling here that a migration is very often the transcendental experience that creates a sense of community and a separate identity. The awareness of a shared experience of pilgrimage provides the common bond to hold together very different peoples under a single religion, for example (Anderson 1983, Turner and Turner 1978). The 'Long March' of Mao Zedong, or the 'Salt March' of the Mahatma Gandhi, also provided 'epic dimensions' to the development of modern China and India. Migration, and circuits of mobility, are crucial in nation-building and in the articulation of the state. Elsewhere, I have shown how the mobility associated with institutions such as the military and education can in effect 'build the nation' (Skeldon 1990) and this is a theme to which I will return in the Conclusion of this book.

Important though these European and national migratory experiences undeniably have been in South African history, it is impossible to understand the recent development of that country without an appreciation of the economic base and the country's position as the world's leading supplier of gold. Gold has been, and is, the leading industry in South Africa, and its exploitation was only possible through the development of a widespread and persistent system of circular black migration. Crush, Jeeves and Yudelman (1991) argue at the outset of their book, *South Africa's labor empire*, that the mining of such deep-lying low-grade deposits of gold ore was only possible because of the state's ability to organize a regular and massive supply of cheap labour. Without the labour system, the state would not have been able to exist. The maintenance of this system of circulation of black rural labour into the heart of a settled white urban economy indeed became 'the keystone of apartheid' (Lemon 1982: 64).

The northward push of the labour frontier into adjacent British and Portuguese colonies from the 1920s through to the late 1960s has been well documented elsewhere (Crush, Jeeves and Yudelman 1991) and only a bare summary of the main arguments is given here. The system of labour recruitment on a short-term basis suited all parties. The employers received a regular supply of largely inexperienced, unskilled labour, for which low wages could be paid. The government of South Africa avoided the establishment of a permanent black urban population, which accorded with its racial policies at that time. The donor colonies benefited from the cash brought back by returning workers as did local chiefs, who also profited from supplying labour to brokers in the first place. Finally, for the peasants themselves, work in the mines was a fairly rational alternative, given the generally harsh conditions in which they lived their lives. At the end of their contracts, the peasants returned for long periods to continue their rural activities, and the mines constantly had to tap new areas to be able to guarantee a steady supply of labour.

From about 1970, however, fundamental changes to the system saw a transition from the remarkably long-established and stable circulatory labour migrations towards forms of mobility that appear to have their counterparts elsewhere. As in the case of China, unique forms of mobility are being transformed through the expansion of a global, free-market system into more general patterns. The freeing of the gold price from the late 1960s saw a sharp rise in its value which permitted companies, in the face of militant labour pressure, to improve real conditions for the miners for the first time in many years. Changing technology in the mining industry increased the demand for more highly skilled labour, thus pushing companies into encouraging their labour to return as quickly as possible from their home areas to avoid the expensive need constantly to train a fresh labour force (see also Lemon

1982). The rise of other manufacturing and tertiary activities in the urban areas also spread the demand for a stable supply of workers. These factors eroded the infamous pass laws, which prohibited the settlement of blacks in urban areas, and from the late 1970s Africans were permitted to acquire long-term leases on houses in urban townships and later to purchase them directly (Omer-Cooper 1995: 845). Finally, the independence of the African nations to the north meant that South Africa had to rely on potentially hostile states for its labour. These states, too, faced a dilemma: whether to deny cooperation with a regime that they found an anathema but lose millions of dollars in remittances; or face guilt and international opprobrium for supporting economically a discredited regime. Malawi, for example, suspended recruitment in 1974 and ordered all 120,000 Malawians home from South Africa (Crush, Jeeves and Yudelman 1991: 103). The South African state thus retreated from its northern source areas and came to depend more and more on domestic and local areas. In 1960, almost half of the 375,614 workers came from outside South Africa, Lesotho and Swaziland; by 1989, all but about 15 per cent of the 421,885 workers came from within that area.

The final legal demise of apartheid came in 1991 and, three years later, multiparty and multiracial elections were held. This dramatic shift in the internal politics of South Africa was, to a large extent, the product of the changing economic base, which had resulted in massive rural-to-urban migration of Africans and the stabilization of these populations. While migration in itself did not cause the movement away from apartheid and towards a more democratic society, it was an important contributory factor in facilitating these changes. Unlike the areas of population concentration in this development tier in Asia, but like those in South America, there is no overall labour shortage. On the contrary, unemployment is one of the major problems facing the new democracy with anywhere between 18 and 29 per cent of the economically active population out of work (Katzen 1995: 856). Yet the new political changes have encouraged renewed migration from other African states, despite the high unemployment in the country. Much of this movement is clandestine, much is fleeing deteriorating conditions in the source areas, to be discussed in Chapter 7, and some is of skilled labour, doctors and teachers, for which there is a demand in parts of South Africa (Stalker 1994: 236).

There are still major problems, both economic and political, to be overcome if South Africa is truly to become a dynamic core in southern Africa. Its well-developed infrastructure and resources, not the least of which is an abundant supply of labour, are powerful forces to promote its position. Internal political stability and the expansion of a market, both domestic and external, with sufficient purchasing power to attract investors are still in the balance. These will need to be assured before we can say that the concentration of population in South Africa is truly leading to a rapidly expanding part of the core economies.

The second area where development has been so fundamentally affected by migration is in the state of Israel. The Jewish peoples have lived in diaspora for some 2000 years and the creation of the state of Israel in 1948 was an apparent prophetic return to the homeland in Palestine. The idea grew out of the disintegration of the Turkish Empire during the First World War and the establishment of a British Mandate in 1923. At that time, the Arabs made up 92 per cent of the population of around 700,000 in the area under the Mandate, and they controlled 98 per cent of the land (Little 1995: 512). Jewish settlement proceeded slowly and in the face of considerable Arab resistance until the 1930s, when the first major wave of Jews arrived, escaping persecution in Europe and bringing capital to establish commercial and industrial enterprises. By the end of the 1930s, however, there were no more than 175,138 Jews in Palestine (Shafir 1995: 407). The next major wave occurred after the end of the Second World War when thousands of survivors of the Holocaust made their way to Palestine. By the beginning of 1946, the number of Jews had increased to some 608,000; it had further risen to 650,000 by May 1948 when the British were forced to terminate the mandate and the state of Israel was created. In late 1947, there had been more than 1.3 million Arabs living in the whole of Palestine and they made up more than half of the population of the area allocated by the United Nations partition resolution to the Jewish state (Adelman 1995). In the violence leading up to the partition, some 400,000 Arabs were evicted from their homes and forced to become refugees in neighbouring countries. Thus, the gathering-in of the exiles in Israel produced an exodus of the Palestinians in a new diaspora that has expelled more than half of them beyond their homeland. Of the 5.4 million Palestinians in 1991, only 2,460,000 lived within the area which had been under the British Mandate and, of these, 655,000 could be classified as refugees in the West Bank and Gaza, and another 300,000 stateless in Gaza (Adelman 1995; see also Abu-Lughod 1995). The diaspora occurred in three main waves: around the formation of the state of Israel in 1948, after the Six-Day War in 1967 and associated with the Israeli invasion of Lebanon in 1982. In total, there were well over 1.2 million movements out of Palestine during the formation and securing of this potential core area.

The new government in Israel, from the late 1940s, embarked on a programme of immigration 'unparalleled in history'. The

Jewish population doubled within three years and went on increasing rapidly, with the two-millionth Jew arriving in May 1962 and the three-millionth early in 1972 (Little 1995: 513). By 1990, the total population, including East Jerusalem, numbered almost 5 million, of whom more than 30 per cent were immigrants (United Nations 1995b). Immigration has continued to the present day and was reinforced by the changed policies of the then Soviet Union, which, in 1989, made emigration easier for Jews. Between then and the end of 1995, some 609,000 people moved from the republics of the former Soviet Union to Israel (*International Herald Tribune*, 29 December 1995).

Unlike the migration to the other areas under consideration in this development tier, the migration to Israel can be seen as a global rather than a regional and local phenomenon. Given the prevailing security situation, local migration from Arab territories is tightly controlled, and it is not inconceivable that the state will virtually isolate itself from its region and become dependent upon distant sources of immigrants and short-term labour. For example, with the closure of the West Bank, Israel turned to a country as far away as Thailand as a source of foreign labour. Some 13,000 Thais were brought in 1995, up from approximately 8000 the previous year, on one-year contracts, renewable for one year only. Rather than an expanding core region, Israel may more truly represent a constrained outlier of the old core. However, immigration from other parts of Asia and Africa figured prominently in the waves of migration of the 1960s and 1970s (Findlay 1994: 55). Recent history has shown that contiguous but separate development tends to be a relatively short-term phenomenon. Given the sharp shifts in the political developments in East Germany and South Africa, and the present directions of the peace process between Israel and the Palestinians, it is not inconceivable to think of an expanding core region in West Asia centred around Israel to incorporate Jordan and the populous Egypt to the south. The current rapid economic growth of Israel certainly favours this scenario. Nevertheless, given the lack of integration of the mobility systems of these areas and current political tensions, such a development appears to lie far in the future and this region will remain a potential rather than a real expanding core.

The final area identified as a potential expanding core region is in India: a broad and discontinuous crescent from the Punjab in the northwest to Madras in the southeast, through the two major cities of Delhi and Bombay, and including most of western India. This area includes the richest agricultural region in India, the Punjab; the national capital, New Delhi; India's greatest entrepreneurial and industrial city, Bombay; and India's link to the dynamic economies of East Asia, Madras. The last of these is important because of the colonial links that established Indian communities, primarily Tamil, in Malaya and Singapore. Their legacy facilitates the penetration not only of Singapore capital in particular but also of Japanese investment, often through Singapore-registered companies.

India, like China, the other global demographic giant, is experiencing pronounced rural-to-urban migration. Unlike China, however, there were no periods of fluctuation when migration was controlled, restricted and reversed, and then released to produce the present wave of short-term and long-term movement. Excluding the huge movements associated with partition in 1947, when between 15 million and 17 million were displaced within two years, and the conflicts that resulted in the creation of Bangladesh, when 10 million refugees moved, there has been apparent overall stability in the population movements within India, with gradual shifts in composition and direction (Skeldon 1986). The incidence of short-term migration appeared to slow down in the 1970s compared with earlier years, while more long-distance, more urban-oriented flows increased. There was also greater participation of women in the flows.

Like China, if in a perhaps less spectacular way, India embarked upon economic reforms from 1991. Foreign investment has been encouraged and areas once closed to the private sector have been opened, but the process of reform has been slow, though progress has been made in certain sectors (*The Economist*, Survey of India, January 1995). It is too early to assess the impact that these reforms may have had on population migration in India but, almost certainly, the impact will be greatest in the western part of the subcontinent, where the majority of privately owned smaller light industries are located. These are precisely the type of activities that will attract female labour at the lower end of the labour spectrum and require high-level skilled labour at the other, in situations similar to those in southern China, Malaysia and Thailand. However, given the recency of the reforms, and political incertitude within India regarding commitment to their full implementation, it is perhaps best to leave these western Indian areas as potential rather than actively expanding cores.

The restructuring core

Until the 1980s, what used to be known as the Second World, consisting of the Soviet Union, eastern Europe and their socialist allies in Africa, Asia and Latin America, could be considered a separate bloc. There was little trade with the First World and, in terms of population migration, the borders were essentially closed, with contact limited to official visits and people fleeing towards the west. Only Poland kept a relatively open border with western countries and, between 1959 and 1979, some 795,000 chose to leave, going primarily to the then West Germany and to the United States (Rhode 1993: 237).

Within the bloc, an even development was a basic tenet of the centrally planned economies, although a core in European Russia has always dominated politically, and gradations in regional economic function could be identified away from that core. For example, migrant labour, which in many people's eyes is the symbol of the capitalist system, was brought into the European socialist core from peripheral areas in Asia, Latin America and Africa (Chapter 6). Since 1989 and the break-up of the Soviet Union, significant movements of population have occurred, both internally and internationally, as the economies have struggled to adjust to a global system. The original core areas of European Russia and eastern Europe are undergoing rapid transformation to adapt to this wider context and are best considered as restructuring cores rather than expanding or potential core areas.

Although the ideal might have been an even development, the reality was very different. Despite controls on the growth of cities, the Soviet Union through most of its history was characterized by very pronounced rural-to-urban migration and the rapid expansion of the largest cities (Ball and Demko 1978, Rowland 1983). The urban population increased by 80 per cent between 1927 and 1958 and even as late as 1979 over half of the populations of Soviet cities were made up of migrants (Yanitsky and Zaionchkovskaya 1984: 471–2). The trend over the last decades before the break-up was for an increasing concentration of population in northwestern European Russia, with distinct south-to-north and east-to-west flows (Rowland 1993). Although the overall mobility rates in the former USSR appear to have been considerably lower than those in the west (Mitchneck and Plane 1995: 21), the centralization of flows in the largest centres of political and economic power was similar. Although there was considerable variation between the countries of Eastern Europe, there was a common pattern of higher rates of internal migration during the 1950s, followed by a gradual slowing, particularly as rural migrants to the cities no longer returned to their villages (Compton 1976). The more industrialized countries such as East Germany experienced rural depopulation.

The shocks associated with the break-up and the opening-up to the west engendered a whole series of changes in the mobility system, not the least of which appears to have been a slowing of urban growth and even a reversal of rural-to-urban movements

(Mitchneck and Plane 1995), partly as a result of continuing state controls on mobility and partly as a result of deteriorating conditions in the cities themselves. A greater shock, however, came from the exodus of people towards the west, particularly to Germany and to Israel. In 1989, some 234,994 left the republics of the former Soviet Union, primarily from the European core areas (Rowland 1993: 156). In 1990, 184,600 Jews alone left the USSR (out of a total of around 450,000), 39 per cent of whom went as 'scientific and academic workers'. Their leaving was a factor in the later decline of these economies (Rhode 1993: 239).

Hence, we have seen a period of intense concentration of population followed by a slowing of migration. Since 1989, there has been pronounced movement to the west out of the most developed parts of the ex-Eastern bloc, a slowing in urbanization and a reversal in migration to the cities in certain areas. From a core-type pattern of migration familiar to us from earlier discussions, the ex-Soviet bloc has taken on distinct forms of movement that might suggest a turnaround in mobility brought about by the undermining of the economic unity of the bloc and economic stagnation. A reversal in the pattern of migration and its relationship with development will become a major theme in the very different context of the resource niche to be discussed in Chapter 7.

Whether the cities of eastern Europe and European Russia can regain their role as poles of attraction and economic dynamism, which are the features or potential features of the areas discussed earlier in this chapter will depend entirely upon future political developments. In common with all the parts of this development tier, political stability and the establishment of institutions for the smooth transfer of power remain the critical issues in whether they become core areas in a truly global system. What does seem clear is that, in the short term, ethnic factors will be significant in future movement: ethnic Germans from eastern Europe and from the lands of the former Soviet Union moving back to Germany and ethnic Russians in other republics back to Russia. Both are the legacy of European and Slavic eastern expansion. The eastern settlement of Germans into Slavic lands dates from the twelfth century in what is now eastern Poland and from the sixteenth century in Russia and further out and, at the time of the great transatlantic movements, Russians were moving eastwards towards Siberia. Some 10 million moved into Siberia between 1815 and 1914 (cited in Baines 1991: 11). There are some 25 million ethnic Russians in the 'near abroad', almost half in the Ukraine, and an estimate of 5 million returnees to Russia over the decade of the 1990s appears probable (Dunlop 1993). Ethnic factors are likely to re-establish the concentration of population in Russia, which may rekindle development, as in most areas of inmigration, particularly if the refugee mentality, discussed in Chapter 4, can provide a stimulus. The new trends in eastern Europe are well detailed in Tomasi and Miller (1992).

Discussion

It is impossible to predict accurately which areas – if any – are likely to be among the next more developed parts of the world. Nevertheless, it is possible almost physically to witness the expansion of development out of core areas in certain locations in East and Southeast Asia and southwards from the United States. To these areas have been added a few key locations that are outliers from the core but that would seem to have the potential to become pivotal development areas for their respective regions. All are centred around major urban areas that are, if not global cities, at least major centres of attraction for local (with the exception of Israel) and regional flows.

Population concentration through rural-to-urban and urban-to-urban migration up the hierarchy characterizes the population movement in this development tier. The establishment of light industries with a strong demand for female labour is also a trend that typifies much of this tier. There are differences, too, some of which have been brought out in the discussion, but among which unemployment is significant. There is a range from areas of low unemployment, as in parts of East Asia, to areas of high unemployment as in southern Africa; despite these differences, all areas are targets of considerable inmigration. Relative to the alternatives available in the areas of origin, the often difficult conditions in the urban destinations in this tier arguably provide a better life. All the evidence suggests that migrants have higher labour force participation rates than non-migrants in major urban areas in Asia and Latin America (reviewed in Skeldon 1990: 161; also Nelson 1979). Labour markets are highly segmented, with migrants from particular areas controlling access to particular occupations. Personal networks channel people into specific activities so that there is no queueing within a single anonymous labour market. For example, in Singapore, secret societies and associations protected regional and speech group monopolies of particular trades and occupations from the late nineteenth century (Cheng 1985). The informal sector, in particular, offers ample scope for a whole range of activities to be developed in construction, manufacturing and services (De Soto 1989). Despite their unofficial status compared with the formal sector, access to informal activities is not easy and is, in fact, tightly controlled through kin, ethnic or place of origin groups. Migrant associations, or groupings of people from the same village or district of origin, also facilitate the incorporation of new arrivals into the city (Skeldon 1976).

While population concentration may be the principal stamp on patterns of migration in this development tier and all centres are areas of strong net inmigration, there has also been outmovement, primarily to countries in the old and new cores. For example, settlers move from more developed parts of China and India to the United States and other parts of the old core. The 'new' migration out of China to the developed peripheral areas of Hong Kong and Taiwan, to which reference was made in Chapter 4, originates primarily in southern China, although numbers are increasing from the Shanghai area and other cities with major universities (Skeldon 1996). The migration from India, since the 1960s, has been characterized by the highly educated. Although the range of places of origin within India has been wide, there has been a clear bias towards Gujarat and the Punjab (Helweg 1991, Tinker 1977; see also Clarke, Peach and Vertovec 1990). These migrants maintain close contact with their home areas, in effect binding them to the more developed parts of the world in circuits of human mobility. Other movements out of this development tier have included skilled temporary migrants moving from Malaysia to other rapidly growing parts of Asia; white South Africans moving to old core areas rather than face an uncertain future in a multiracial country; European settlers returning home from Argentina and Brazil; and Russian Jews and other members of the intelligentsia from the western part of the former Soviet Union moving to Israel and to Europe.

All of these circuits of mobility reinforce the linkages between this development tier and core areas. Whether the expanding core can make a transition to a stage when it, too, can export capital and migrants further out into less developed areas will ultimately depend not just upon its ability to move towards higher levels of capital accumulation and technologies of manufacturing, but on political stability. Major question-marks still hang over the long-term political direction of several parts of this development tier. Often, channels for mass participation are either not well developed or under stress through religious or ethnic divisions. The concentration of populations has emphasized wealth differentials between workers and managerial groups and, although worker pressure has achieved some successes in both northern Mexico (Sklair 1993) and South Africa (Crush, Jeeves and Yudelman 1991), much remains to be done. In some areas, notably southern China, there are few means for the migrant population to achieve a more equitable distribution of wealth. Undisturbed continuous development is not guaranteed in some of the areas of most rapid growth in this development tier. Although there is immigration of the highly skilled on short-term contracts from core areas into this expanding core tier, the vast majority of the immigrants come from the tier that I have identified as the labour frontier, and it is to these areas that we now turn.

Chapter 6

The labour frontier

The fourth development tier, as its name implies, is conceptualized as being among the main areas of origin of migration to the principal destinations. This does not mean that there are no destinations for migration within the labour frontier: there are many. Again, the artificiality of drawing clear boundaries between spatial regions is highlighted. It could be argued that some of the areas included in this development tier could be more appropriately considered within the rapidly expanding core tier. The Jakarta extended metropolitan area, or central Chile, would perhaps be two of the most obvious examples. Yet, no part of the region as defined in Fig. 2.2 appears to have the pulling power, either at present or in the immediate future, to attract large numbers of migrants from more than a local or national hinterland. On the contrary, this development tier is dependent upon migrants moving outwards to the expanding core, to the cores themselves, and to parts of the final development tier, the resource niche covered in Chapter 7.

The migration in and from the labour frontier is much more difficult to describe simply. The patterns of migration in core areas (Chapters 3 and 4) were characterized by increasing immigration and trends towards the deconcentration, or at least the stabilization, of highly urban populations; the expanding core areas (Chapter 5) were typified by intense concentration of population in cities. The patterns of migration in the labour frontier are much more complex. There is movement within the rural areas themselves, there is concentration in large and small cities within the tier, and there is movement directed outside the tier, not always perhaps the most important in terms of volume but the most significant from the point of view of development.

The labour frontier covers large parts of central China along an approximately north-northeast to south-southwest axis; it includes the densely populated archipelagic zones of Southeast Asia in Indonesia and the Philippines; it incorporates northeastern Thailand and much of Indochina and Burma, most of eastern and northern India, Bangladesh, and much of Pakistan. In Africa, it includes a zone along the northern coast from Morocco to Egypt and also a group of countries in the south, including Namibia, Botswana, Lesotho and southern Mozambique. The tier includes much of eastern Europe and Turkey, and western Russia and the

Ukraine. Finally, in Latin America, it covers most of Central America, including parts of Mexico, the western parts of the Andean republics and northern and coastal Brazil.

The evolution of the labour frontier

Just as we could see the core expanding in parts of East and Southeast Asia and in North America, so too the boundary delimiting the labour frontier has moved and is continuing to move. Some twenty and more years ago, the labour frontier for the European core would have lain much more in the southern parts of that continent itself in Portugal, Spain, Italy and Greece. However, as we saw in Chapter 3, these countries have gone through a transition from labour exporter to labour importer, and the labour frontier today lies further to the south and east. Italy has had a positive migratory balance since 1972, Spain and Greece since 1975 and Portugal since 1981 (King and Rybaczuk 1993: 176). These transitions, however, as seen in Chapter 4, are seldom neat, unilinear events but gradual trends moving along several paths towards a convergence. There is still labour migration out of countries in southern Europe, and from Portugal in particular, into the old core. The volume of migrant remittances for Italy, Greece and Portugal is still significant in absolute terms, surpassing those for 'classic' labour-frontier countries such as Turkey, Egypt, Morocco, India or Pakistan. However, in terms of their contribution relative to GDP, remittances for the southern European countries, with the exception of Portugal, are of lesser importance (see the data in Russell and Teitelbaum 1992). The trend of the labour frontier is away from European sources towards more distant areas in North Africa, Turkey and beyond. These trends are evident in the data on foreign labour in France and Germany (Table 5).

Labour migration from the new parts of the labour frontier did not suddenly begin with post-war European demand: it was the result of a long gradual process. In North Africa, traditional systems of seasonal circulation were modified from the 1860s by colonial penetration, which tapped the densely populated mountain areas for agricultural labour for European farms on the northern coast (MacMaster 1995). Northern Algeria was the earliest and most profoundly affected and, with the extension of commercial agriculture there, the labour frontier was pushed both westwards and eastwards into Morocco and Tunisia. After the First World War, as more Algerians began to go seasonally to France, increasing numbers of Moroccans were recruited to

Table 5.   Stock of foreign population by nationality: France and Germany, selected years (thousands)

	France			Germany		
	1975	1982	1990	1978	1982	1990
Portugal	758.9	767.3	649.7	110	106.0	84.6
Algeria	710.7	805.1	614.2	–	5.1	6.7
Morocco	260.0	441.3	572.7	29	42.6	67.5
Italy	462.9	340.3	252.8	572	601.6	548.3
Spain	497.5	327.2	216.0	189	173.5	134.7
Tunisia	139.7	190.8	206.3	19	25.2	25.9
Turkey	50.9	122.3	197.7	1,165	1,580.7	1,675.0
Former Yugoslavia	70.3	62.5	52.5	610	631.7	625.5
Greece	–	–	–	306	300.8	314.5
Poland	93.7	64.8	47.1	–	91.4	241.3
Senegal	14.9	32.3	43.7	–	–	–
Viet Nam	11.4	33.8	33.7	–	–	–
Iran	–	–	–	–	32.2	89.7
Others	371.5	526.5	710.2	981	1,076.1	1,428.1
Total	3,442.4	3,714.2	3,596.6	3,981	4,666.9	5,241.8

Source: SOPEMI, *Annual report: trends in international migration.* Paris, OECD, various years.

replace them. Thus, along the southern periphery of Europe, the labour frontier gradually expanded in a system of stage migration somewhat reminiscent of Ravenstein's original formulation. For a detailed description of the evolution of the spatial patterns of international migration in North Africa over the more recent past, see Findlay and Findlay (1982).

The colonial roots of labour migration within and from North Africa to France are clear. There was no direct colonial involvement in Turkey or in eastern Europe but there too, external influence was important. Linkages were established by German businessmen and technicians in Turkey after the First World War. Much earlier, from the sixteenth century onwards, throughout much of eastern Europe and Russia, the immigration of German professionals and craftsmen 'made decisive contributions towards modernizing the economy, administration and military forces' (Bade 1995: 131). These skilled migrations were followed and reinforced by colonies of rural German settlers. These early linkages were of importance in accounting for the much later development of labour migration from Turkey (since the mid-1950s), from the former Yugoslavia (in the 1930s, and again since the mid-1960s), and in explaining the streams of movement from eastern Europe following the collapse of the Soviet system.

The labour frontier in other parts of the world has also evolved from pre-existing systems of circulation modified by later penetration of outside influences; some of these were colonial, others of less direct but no less disruptive sources, and still others of mainly indigenous origin. Pakistan, which, as we will see, is an area where labour migration has had a most profound impact, was a highly mobile society with a long history of labour movements, not only under the British, but also traditionally to the Gulf region (Addleton 1992). Traders from areas now in western India and Pakistan have had links with, and established communities in, the states around the Persian Gulf over several centuries and, when the Persian Gulf states were under British Protectorate in the 1930s, workers were sent from the then British India (Gogate 1986: 39). Hence, although the marked movement to the Middle East associated with the oil boom of the 1970s was new in terms of numbers, the framework for much of that migration had been laid long before.

The first great phase of a labour frontier in South Asia, however, dated from 1834, when slavery was abolished in the British Empire, and lasted until 1917. Some 1.5 million indentured labourers were moved to places outside the South and Southeast Asian region and another 6 million were moved to destinations in other parts of that region in Burma, Malaya and Ceylon (Clarke, Peach and Vertovec 1990: 8). These movements established a widespread network of overseas Indian communities whose members have played a key role in societies as diverse as Singapore, Malaysia, Mauritius, South Africa, Trinidad and Tobago, and Fiji. The importance and consequences of these earlier labour migrations have been well detailed elsewhere and the following discussions will focus primarily on the present labour frontier (see Tinker 1974, 1977; and the essays in Clarke, Peach and Vertovec 1990).

Other areas where, like Pakistan, labour migration has become a way of life are Mexico and the Philippines. The case of Mexico was discussed in the context of the expanding core in Chapter 5 but much of that country is a labour frontier supplying workers not only to the main urban areas, particularly Mexico City and the *maquiladora* industries previously discussed, but also to the United States. Beginning in the late nineteenth century, the movement to the United States had reached massive proportions by the late twentieth century with 1,650,000 being admitted legally during the decade of the 1980s and hundreds of thousands more entering illegally. The Philippines has emerged as one of the principal sources of migrant labour in Southeast Asia, with workers in over 130 countries around the world. From a nation of some 69 million, there are an estimated 4.2 million workers and settlers, both legal and illegal, overseas (Yukawa 1996: 1).

The regional origins of labour migration

An overall figure for the number of migrants overseas is not particularly meaningful for the Philippines, or for any other country for that matter, as the origins of the migration tend to be concentrated in certain regions. There does appear to be a tendency, although not necessarily an invariate one, for these areas to be on the periphery of, or at least close to, the expanding core regions. In Mexico, the states of Michoacan, Jalisco, Guanajuato and Zacatecas are among the most important of the traditional sources of migration to the United States (Lozano Ascencio 1993: 66). Most of the legal migration from China comes from areas near to the regional growth centres around the Pearl River Delta and much of the illegal movement is from districts in Fujian Province close to, but not quite in, the growth centres of southern China (Skeldon 1996). The main areas of origin of internal migrants in China itself lie further in the interior, in the densely populated province of Sichuan in particular, but also from those provinces such as Guangxi, Hunan and Jiangsu which lie just inland from the southern and central expanding core areas.

More than half of the workers who have left the Philippines to go abroad came from Manila and the provinces in Luzon close to the capital (Arcinas 1986: 268–9). Workers from Thailand tend to come from the northeastern and northern provinces, although large numbers were recruited in the capital and were domiciled there before going overseas (Chiengkul 1986: 314). Internal migration is often a prelude to international movement. The labour migration from these two countries also seems to have been biased towards those areas with close linkages to an outside power or, in these cases, to the areas where there had been a strong United States presence through air bases and military personnel. When American contractors who had been involved in the construction of military facilities in northeastern Thailand during the Viet Nam war established themselves in the Gulf, they looked to their tried and trusted Thai workers as a labour force (Chiengkul 1986). Almost one quarter of workers going overseas from the Philippines came from provinces where there had been large United States military bases (Cariño 1992: 12–13).

American military bases can have other long-term impacts on migration. They can be a factor in the promotion of settler migration as a result of servicemen marrying local women who, on returning to the United States, then bring in family members. Kuznets (1987) estimated that one out of nine American servicemen based in South Korea at one time would return with a Korean wife, which plants a considerable seed for further chain migration. An even more interesting example of the impact of American bases comes from Ota city and Oizumi-machi prefecture in Japan, northwest of Tokyo. The major American bases that had been established there following the Second World War were handed back to the Japanese government in 1959. Not only did the returned facilities provide the basis for the establishment of new industrial zones but the local population had been accustomed to the presence of foreigners and to the prosperous local market that they provided. Local entrepreneurs actively sought foreign labour from shortly after the departure of the Americans to help with the industrialization. Today, the two neighbouring areas have foreign populations representing 9.4 per cent in Oizumi-machi and 3.2 per cent in Ota city in early 1996, among the highest proportions in Japan. The military can thus play an important indirect role in the expansion of capitalism and the resulting immigration, as well as a direct role in migration through recruitment. These data (from a site visit in March 1996) support the importance of the history of military and political involvement in explaining current international patterns of migration (see Sassen 1988).

Labour movements appear to be most intense from those areas which lie on the interface of the core, or the expanding core, tier and the adjacent parts of the labour frontier. In India more than half of the workers leaving for the Middle East originated in Kerala, immediately to the south of the area identified as a potential expanding core, with the majority of the balance coming from other states in the west of the country (Nair 1986: 70). The way in which the system can evolve in stages may be seen in the case of southern India with carpenters and others involved in the construction industry moving into Kerala from the adjacent state of Tamil Nadu (Nair 1989: 353–6). Whether this movement was in response to skill shortages in Kerala directly caused by the emigration, or in response to increasing demand for workers owing to the boom in the construction industry based upon a flood of remittance money, is immaterial. The end result is a stage-type penetration into the labour frontier. The progressive impact of international migration on internal movements is as clear in India as it is in a different way in China, where the investments, primarily from overseas Chinese sources, and thus the product of previous emigration, are now giving rise to some of the greatest peacetime internal movements in that nation's history.

The case of Pakistan provides interesting comparisons and contrasts. Although the main cities of Karachi, Lahore, Islamabad/ Rawalpindi and Peshawar are disproportionately represented in the flows of workers going abroad, the majority of migrants came from poorer and isolated rural districts in the northern Punjab and North West Frontier Province (Addleton 1992: 95–6). This perhaps atypical pattern, which Addleton contrasts with the evolution of labour migration from Turkey (see also King 1993b: 26), where there was a clear movement from richer to poorer areas, was perhaps due to the highly mobile nature of most parts of Pakistan (rural Sind would be an exception here) and the opening-up of these isolated areas to military recruitment under the British. The military, like education, was not only an important institution in fostering national identity but also significant in initiating circular migration out of areas that had little previous contact with central governments (see Skeldon 1990: 198–201).

The idea that labour migration is generally at its most intense along the interface between the expanding core, or the core areas themselves, and the labour frontier tier may give the impression that the frontier must push outwards as appears to have been the case in Europe or in Asia. Nevertheless, that frontier can retreat, too, as in the case of southern Africa discussed in Chapter 5, where the regions of origin of movement into that potential core have, for political reasons of security and economic reasons of high unemployment, moved back, closer to the potential growth area. There is thus nothing inevitable about an expansion of a labour frontier, and the contraction will be discussed in more detail in the context of the particular conditions of sub-Saharan Africa in Chapter 7.

The recruitment of labour

Common to all parts of the labour frontier is the presence of the labour contractor. To be sure, there is much movement out of the labour frontier on an independent basis, not so much the result of some totally individual decision, but one taken within the context of social networks and information and assistance provided by previous migrants. These provide the basis of the long-standing chain migration interpretation, with migration leading to further migration. The strategies of these independent movements can be understood within the context of risk-minimization strategies, discussed in Chapter 1. The destination is incorporated as an additional niche in the resource base of the domestic economy. A key issue is how the initial contacts are made with potential destination areas in order to allow the chain to proceed. Here, the labour recruiter or the broker plays a key role. Governments of countries in both the labour frontier and the core and expanding core areas (in the case of internal migration these are obviously the same) can be directly involved in labour recruitment as can individual company employers and specialized recruiting agencies. See Castles (1995) for a concise and useful review.

The trade in human beings is virtually as old as history and has ranged from slavery (Chapter 3) to the many forms of contract, formal and informal, that can be found among labour migrants today. Because governments, companies, and legal or illegal contractors have some control over a worker's life for a period of time, contract labour is seen as a form of 'unfree' labour (Cohen 1987, Miles 1987). Since the Industrial Revolution and the need to transfer workers to labour-deficit areas, the labour contractor has been a virtually ubiquitous feature of migration systems, both internal and international, as common in rural areas of late eighteenth-century Britain (Redford 1976: 22) as in early twentieth-century Europe (Castles 1995: 510), the highlands of southern Peru in the 1950s (Skeldon 1990), the South African labour empire since the late nineteenth century, and in China, Pakistan or the Philippines today. As discussed in previous chapters, there has been 'leakage' from these circulatory systems of migration and, in Europe and the United States, the labour migration of North Africans, Turks and Mexicans has led to permanent settled ethnic communities in the destination areas. Any distinction between labour migrants on the one hand and settlers on the other becomes blurred. In post-apartheid South Africa, too, many labour migrants are becoming settlers (Chapter 5).

Over 95 per cent of Thai migrants to the Middle East were recruited through private channels and the cost of obtaining a job was some $US1200, excluding what it might cost to go to Bangkok to wait for perhaps months while the paperwork was completed (Chiengkul 1986: 320). In Pakistan, in the late 1970s, private recruiting agencies were demanding the equivalent of almost $US3000 to find a job; and this cost, as in the Thai case, excluded the airfare. Small wonder then that, although hard data on the prior income level of the labour migrants was difficult to obtain, it was observed in Pakistan that the income of pre-migrant households was certainly above the national average and that 'emigration has been taking place from the relatively more affluent part of the population' (Fahim Khan 1986: 118). However, we have already seen that many, perhaps the majority, of the labour migrants from Pakistan came from the poorer areas of the northwest. The reason for this apparent anomaly appears to lie in the nature of the early migration to Saudi Arabia, which was largely controlled by companies that had been engaged in construction in the northern and western parts of Pakistan, as well as in Karachi, and which transferred their labour to the Middle East once they secured contracts there. Only later did private recruitment companies come to dominate, and then the movement was presumably from more affluent areas. The construction boom in the Middle East fortuitously started just as the large projects in Islamabad and major irrigation works in the Indus valley were coming to an end (Burki

1991: 144–5). The connections between prior internal migration and later international migration through transnational companies or other institutional linkages, as in the case of the military discussed earlier, is again clear. In fact, in the Pakistan case, long-distance migration out of the most isolated areas was, in a large number of cases, initiated through military recruitment that led to later employment in construction.

The rights of migrants

The labour contract system is open to abuse, and since the long campaigns to abolish slavery, attempts have been made to protect the rights of migrant workers. These attempts have mainly been through the United Nations and its specialized agency, the International Labour Organization, and have become associated with more general principles on human rights. Giving rights to migrants who are foreigners causes many governments to be uneasy for fear that such treatment would be a prescription for full settlement and thus a potential change in the nature of their society. The tension between the protection of migrants on the one hand and safe-guarding the legitimate concerns of the state to protect the rights of its own citizens on the other remains one of the great and largely unresolved issues in the general area of migration today (see, for example, Weiner 1995).

Apart from denying any way by which migrants can become citizens, a common policy to deter long-term settlement is to prohibit the entry of wives and children. As soon as family members, and women in particular, participate in population movement, the incidence of turnover in circulation appears to slow as the migrants spend longer at their destinations. As Hondagneu-Sotelo (1994) suggests from her analysis of Mexican migration, men are more likely than women to consider themselves as sojourners and she attributes this to their constant attempts to re-establish their independence. Women are certainly often seen to be the principal means by which settlement is achieved. During the apartheid period in South Africa, the infamous pass laws were particularly applied to women who were likely to be 'a harbinger for the establishment of black families' (Cohen 1987: 165). The settlement of women, and thus the potential that children are born to migrants at the destination, assumes special importance in those countries that award citizenship on the basis of place of birth rather than on descent. In legal terms, the former are countries that follow principles of *jus soli* – of the soil – as opposed to those that follow *jus sanguinis* – of the blood. France, the United States and Canada are examples of the former, while Germany and Japan are examples of the latter. There are nevertheless variations upon this simple dichotomy depending upon the status of one or both of the parents. The United States, however, currently allows the option of citizenship at age of majority to any person born on its soil, irrespective of the resident status of the parents.

Few states accord rights to migrants immediately upon arrival that give them equal legal status with the indigenous population; the exceptions are based upon specific cases of ethnic (Germany) or religious (Israel) affiliation. Discrimination, legal as well as social, exists in countries right across the development spectrum, but it is most severe outside the core tiers, where the rule of law and state institutions are generally more weakly developed. Movement from the labour frontier into the expanding core and, more especially, the resource niche is particularly liable to abuse. This does not mean that all migrants into core areas are accorded protection. They are not, particularly where the movement is illegal or undocumented and thus 'unseen' by the state, as in the case of much Mexican migration into the southern United States or of Chinese into the sweatshops of New York city. Unfortunately, the perceived fear of rising immigration may be leading to an increasing global trend in state discrimination against the outsider (Richmond 1994). Clearly, too, where the movement is within a country, the protection of migrants is not subject to international scrutiny and is exceedingly difficult to monitor. Trafficking in women and children from many parts of the labour frontier to the major regional and global cities is an especially vexed question. The movement of women to the brothels and massage parlours of Bangkok, initially mainly from the north and northeast of Thailand but increasingly from further afield in Burma, Laos and southern China, is just one example of this trade out of the labour frontier.

Characteristics of the migrants

As virtually all migration studies have shown, the majority of those who move are young adults. Apart from this generalization about the characteristics of those who move, virtually every other statement requires some qualification. Nevertheless, some trends can be identified. In most cases, migration out of the labour frontier is initially dominated by men, but with increasing participation of women over time to the extent that certain flows become female-dominant. The shift in the sex balance occurs primarily in the movement up the development hierarchy, whereas that from the labour frontier out to the resource niche is still dominated by men. The nature of the industrialization in the rapidly expanding core and the demand for domestic servants (Chapter 5) largely explain this pattern, although in some parts of the labour frontier, in Latin America and much of Asia, women can be released from the rural labour force much more easily than in other areas. Throughout the islands of the Pacific and in much of sub-Saharan Africa, for example, women undertake the key agricultural tasks – these are areas of 'female' farming (Boserup 1970). Men can move from these areas on a more long-term basis without making a significant impact on rural production, and it is in these two geographical regions that men have tended to dominate the patterns of circulation. That said, however, for a variety of reasons, discussed in Chapter 7, women in both these areas are now migrating independently in increasing numbers.

Rarely is it the poorest from any community who move. The cost of transport, the cost of establishing and maintaining oneself at a destination and, perhaps most significantly, the cost of the labour contractor virtually ensure that the migrant must have access to resources. Even under conditions of considerable physical duress, as in the case of Ireland in the nineteenth century, the poorest tend not to move. In that previous labour frontier, now in the old core (although perhaps a case for Ireland remaining part of the labour frontier can still be made – see the Conclusion), emigration out of Ireland in the eighteenth century was small compared with that of the nineteenth century. Yet, people were more destitute in the earlier period compared with the later period (Miller 1985: 131). In Italy, during the phase of migration to North America, there was 'a level of misery below which emigration was simply impossible' (Arlacchi 1983: 183). It is not poverty that makes people move but 'change' (Miller 1985: 131), which can be interpreted as development in a broad sense. Such change is engendered by increasing knowledge of opportunities available elsewhere and, to a large extent, is a function of prior movements; it was the relatively better-off, with their higher education and greater awareness, who could respond to that information. The general conclusion of studies of labour migration, and of migration in general, is that the migrants come from the wealthier sections of their local society, even if that wealth has to be seen in relative terms. Poorer migrants, and migrants from poorer areas, can also be involved where they are facilitated by institutional factors such as the military.

The impact of migration on the labour frontier

Among all the diversity in this development tier there is the common theme of dependence upon destination areas. Although there is long-term and virtually permanent migration out of and within parts of the labour frontier, one of the principal character-istics of the population mobility of this development tier is circula-tion. This may be either long-term or short-term and it brings to the fore two apparently opposing trends that are often at the centre of the analysis of the relationships between migration and develop-ment. As we saw in an earlier discussion, migrants include the best-educated – even if 'best' is necessarily relative – and most dynamic members of their communities. Is the loss, even the temporary absence, of these people in some way negative for the development of the communities of origin? On the other hand, the migrants seldom sever their ties with their home areas. Many of those on labour contracts choose to return upon completion of the contract but those in more long-term employment in urban areas also return periodically and may settle back home after years away or upon retirement. The skills learned away from home that returning migrants take back and, perhaps more importantly, the money and goods that they send and bring back have to be counterbalanced against any losses that their absence from the home area may have created.

Skill transfer, or more accurately the transfer of new attitudes, was important in the case of the return of students in East Asia (Chapter 4), and remittances within the core areas can also be significant as, for example, in the case of state remittances in the form of pensions (Chapter 3). However, although skill losses and the flows of remittances are by no means restricted to the labour frontier, their relative impact is almost certainly greater in this tier, and also in parts of the resource niche (Chapter 7). In the core and expanding core areas, remittances are just one flow among many and are dwarfed by the volume of foreign investment and trade and will rarely amount to more than 1 per cent of foreign exchange earnings. In the labour frontier, and in parts of the resource niche, the relative impact that migration can have on development is much greater and skill losses, skill transfers and remittances can indeed be significant. For example, fully one quarter of the foreign exchange of Bangladesh, a country of some 115 million, is esti-mated to come from migrant remittances. See United Nations (1995b) for a useful global summary. Remittances are not only 'central to the links between migration and development' (Russell 1992: 267) but their role in that development is also controversial as to whether the flows are positive or negative for the receiver. The consequences of the migration for the labour frontier areas will be assessed in terms of their demographic, economic, and social and political impacts.

Demographic consequences of migration

Although the absolute numbers leaving parts of the labour frontier are impressive, a stock of over 4 million from the Philippines in 1995, and a flow of some 130,000 from Sichuan Province of China alone between 1986 and 1990, for example, they are seldom significant enough to have more than a local impact on total population growth. The emigration from the Philippines represents around 6 per cent of the total population, which is much less, thus far at least, than the proportions of emigrants from European countries early this century (see Abella 1992: 23). The majority of countries in the labour frontier, with the notable exception of eastern Europe, still have relatively high rates of population growth or, more appropriately, of the growth of their labour forces (Annexe Table 1). Abella (1992: 26) has estimated that the emigration from the Philippines of settlers and contract workers in the early 1980s reduced the growth of the resident population by some 140,000 per annum, but this represented only about 12 per cent of the total annual increase. The emigration from Sweden from 1850 to 1930 accounted for about one third of the natural increase over this long period (Mosher 1980a: 401). However, among smaller nations in the labour frontier in Latin America, such as Paraguay and El Salvador, the emigration rates appear to 'have been high enough to offset natural increase rates' and the departure of 150,000 women of childbearing age appears to have been an important factor in the declining birth rates in Colombia (Díaz-Briquets 1991: 185–7). Emigration can also cause population decline among the small populations in the resource niche (Chapter 7) and, as will be seen below, a decline in parts of the present labour frontier in eastern Europe.

Rather than significantly affecting the overall rate of population growth, overseas migration may have had a greater impact on internal distribution, again suggesting the linkages between internal and international migration. For Pakistan, 'emigration to the Middle East appears to be the most potent factor that seems to have lowered rural-urban migration in 1972–81' (Abbassi cited in Addleton 1992: 171). Movements that would have been directed towards local cities have gone overseas instead, international migration thus substituting for internal movement. Yet, vacancies that were created by urban residents moving to the Gulf may have been filled by rural migrants. In this case, international movements stimulated internal migration, even if there had indeed been an overall slowing in urbanization. Thus, the linkages between internal and international moves are complex and are likely to change over time.

As we have seen in earlier discussions, the migration is

selective of specific age groups and of specific areas. Its local impact may thus be considerable but more in terms of a skewing of sex ratios than of a decline in population. This imbalance has perhaps been more important in certain Islamic countries, such as Pakistan, where the movement of women has been actively discouraged, but that of men promoted. In the 1970s and 1980s, 'as many as a million wives were separated from their husbands for long periods of time' and, at the level of particular villages, almost half of the wage-earning men were overseas or in cities (Addleton 1992: 156, 175). Conversely, in other societies where there is a high demand for female labour in destination areas (Chapter 5), villages may have large numbers of their young women living outside the labour frontier. In most of these cases, the young men will also have migrated, although there may be bottlenecks in the marriage market with some men not being able to find a wife.

Rather than the impact of migration on the sex ratio, a much more long-term problem may be the impact of an unbalanced sex ratio on future migration. The imbalance is caused not so much through selective outmigration but by a strong preference for sons where there are social and medical practices available to bias the outcome as in East Asia and parts of South Asia. A rising imbalance in the sex ratios in China, South Korea and parts of India is causing concern as the implications of male-dominant societies for the future economy and society are unknown (see Park and Cho 1995, Zeng et al 1993). These imbalances affect not only areas in the labour frontier but in the more developed parts of these countries in core and expanding core areas too. While migration is certainly not the only possible consequence of these imbalances, the importation of women for marriage, by trafficking or broking, from areas within the labour frontier and beyond may indeed become a significant flow in the future. Thus, we may see a major extension of 'mail-order' brides, currently supplying old core countries, if on a fairly small scale, with marriageable women, mainly from Southeast Asia. The 'idea' of Australians marrying Filipinas within Australia, and of Filipinas marrying Australians within the Philippines, diffused progressively outwards from the largest cities in both countries from the mid- to late 1960s (Jackson 1989), although this was not the principal flow from the Philippines to Australia. See Cahill (1990) for a discussion of the marriage 'squeeze' in Japan, Australia and Switzerland owing to an excess number of males in the marriageable age groups, particularly in the rural areas of Japan and Switzerland, and the resultant importation of Filipina women.

Economic consequences of migration: the labour market

Of more immediate consequence for development than the above long-term implications are the economic impacts of migration on the labour frontier. These can be subdivided into two general areas, the first related to the above demographic aspects, the labour market, and the second to flows of money.

Given the rapid rates of growth of the labour force throughout much of the labour frontier, governments have been concerned about rising domestic unemployment and have seen overseas migration as a means of alleviating the problem. Yet, this goal has seldom, if ever, been achieved. As in the case of population growth, the numbers of labour migrants are generally small compared with the total labour force. For example, the number of contracted labourers from India in peak years during the early 1980s was equivalent to only 1.7 per cent of its unemployed workforce (cited in Stalker 1994: 115). In countries such as Turkey, where workers overseas represented a considerably larger proportion of the labour force, about 6 per cent, the unemployment and underemployment rates had not fluctuated significantly since before the migration had begun, 'making it hard to link emigration patterns with trends in unemployment' (Martin 1991: 51). Again, unemployment in the Philippines has been so high that the very substantial annual labour migration of around 430,000 has not made a significant impact on unemployment and is 'at most a relief to the economy and will not lead to structural change ... [although] it has probably meant the difference between collapse and survival' (Vasquez 1992: 47). In Bolivia, emigration removed only about 9 per cent of the projected increase in the labour force between 1960 and 1970 and in Montevideo, Uruguay, unemployment increased from 8 to 13 per cent in the mid-1970s when emigration to the core area in Argentina contracted (Díaz-Briquets 1991: 187).

Any attempt to relate the number of workers to aggregate labour markets, or to the number of unemployed, is not going to be particularly meaningful. The majority of those going overseas were employed at the time of their move and came from particular segments of the labour market. One third of those leaving Turkey in the mid-1960s were classified as skilled migrants; they included among their number 5 to 10 per cent of the stock of plumbers and electricians and 30 to 40 per cent of the stock of carpenters, masons and miners (Martin 1991: 52). Similarly, labour migrants from Pakistan included significant proportions of particular skilled production workers, with over half the number of carpenters, masons, plumbers, nurses, accountants and engineers employed abroad as employed domestically (Addleton 1992: 170–1). The migration creates vacancies which the unemployed or new entrants to the labour force might fill (Martin 1991: 52) but to fill the vacancy with people possessing similar expertise is more problematic, and the real concern is thus more about skill loss than any direct or indirect impact that the migration might have on unemployment. Yet, even with those significant skill losses, albeit temporary in the majority of cases, no impact on production output was observed in Pakistan, Turkey or the Philippines (Stahl and Habib 1991). The periods of maximum outmigration in Pakistan coincided with periods of high rates of industrial growth (Addleton 1992: 181). Migration from the labour frontier, like the movement from the expanding core areas at an earlier stage of development, does not appear to have a major negative effect on the labour market. The principal impact of migration on development revolves around the flows of capital back from migrants to the countries and communities of origin in the form of remittances.

Economic consequences of migration: remittances

In common with so many variables associated with migration and development, remittances are notoriously difficult to measure accurately. The available data often refer to the amounts of cash transferred through official channels, and it is known that much, perhaps most, is sent through informal channels or brought back when the migrant returns for visits or for good. Remittances, themselves, can consist of several types of cash flow: 'labour income', as well as 'workers' remittances', and the value of goods brought or sent back. In addition, there is the important dimension of human capital, which is even more difficult to measure objectively: the skills learned overseas and brought back by returned migrants. The most comprehensive review of the factors controlling the volume of remittances is still Russell (1986), but see also Russell (1992) and Russell and Teitelbaum (1992). The effects of remittances need to be considered at the macrolevel of the national economy, as well as at the microlevel of individuals, families and local communities. While the consequences of remittances from internal migration at the microlevel may be similar to those from international movements, except perhaps in terms of size, internal transfers have few macrolevel impacts and no effect on the growth of the money supply.

Russell (1992) estimated that the global volume of observable remittances in 1990 was $US71.1 billion, making it second only to oil in terms of value in international trade. The relative importance of remittances clearly varies by country and by area within those countries. In the case of Pakistan, official recorded data show that remittances approached $US3 billion during the peak year in 1983, and were maintained at over $US2 billion per annum throughout most of the 1980s. For India, the figures were generally slightly higher, and for Egypt they were well in excess of $3 billion per annum (data cited in Russell and Teitelbaum 1992); for Mexico, an intermediate estimate is around $US2.2 billion per annum (Lozano Ascencio 1993). In Asia, remittances made the greatest relative impact on Pakistan, representing almost 9 per cent of GDP in the mid-1980s, and they were 'an important factor in allowing Pakistan to sustain the highest growth on the South Asian subcontinent through most of the 1970s and 1980s' (Addleton 1992: 123). States have come to depend upon remittances as a source of foreign exchange and, for many countries (including Portugal and Greece), the value of remittances exceeds 10 per cent of their foreign exchange earnings. China, a relatively late entrant into the labour migration market, clearly sees, at the central government level, its vast population as a resource to be invested overseas in the way that the Philippines has been able to do so successfully (Fang 1991).

Important though these aggregate macrolevel considerations are, the critical issues revolve around what individuals, rather than countries, do with the remittances: whether they are used primarily for consumption, which may lead to increasing imports and the loss of foreign exchange and then be negative for development, or whether they are used for savings and investment, which are productive and hence positive for development. As with most cases where we try to simplify into binary opposites, the real situation is more complex and the two interpretations are not necessarily antithetical. Studies on the use of remittances sent home by settler migrants in Britain from the Punjab, in India, have suggested a three-phase succession: providing for the family, conspicuous consumption and business investment (cited in Addleton 1992: 152). Because labour migration has occurred over a much shorter time period than the settler movements, these phases have tended to become compressed and less distinct, although all three types of usage have been observed. Conspicuous consumption, which is often viewed in a negative light, in fact stimulates demand for local production, both agricultural and industrial, and for local services, thus generating employment. Again, in the case of Pakistan, the remittances helped to stimulate the small-scale manufacturing sector within the country (Addleton 1992: 121–7). Clearly, in countries without a basic manufacturing capability, remittances may indeed increase dependence on external supplies, but they can, in other cases, foster indigenous development directly and indirectly.

The consensus from microlevel studies on the use of remittances indicates that 'the average migrant worker spends his money prudently' (Gunatilleke 1986: 129). Poor households do benefit and the physical quality of life improves. Inequalities may increase between households that send out migrants and those that do not, but Addleton (1992), in his assessment of labour migration, has argued that remittances spread benefits to areas and to groups of people well beyond the limits of the traditional elite, who tend to control so much of development funds. In effect, labour migration 'undermines the centre', in Addleton's words, as it has occurred largely with the state as an interested but ineffectual observer and has extended development away from the large centralized government projects typical of so much planned development. Thus, labour migration can be a powerful force not only for economic change but also for social and political transformation.

A distinction must be made within the labour frontier between what has been seen as a labour reserve and the system of labour migration described thus far. In a labour reserve, a labour force is maintained for capitalist production in which the costs of reproduction of the labour force are borne by the areas of origin, with labour subsidizing capital. In part, this situation describes most areas of the labour frontier, particularly in its initial phases of development, though the labour empire in southern Africa, outlined in Chapter 5, can be taken as one of the best examples of the labour reserve. Any system, however, that depends upon the short-term circulation of labour is transferring the costs of the production of the next generation of labourers back to the areas of origin. The costs of raising of children, production of food, housing and so on are borne by the domestic economy, not the capitalist economy (Meillassoux 1981). The critical aspect in the maintenance of the system, and a basic contradiction, is that money must not be allowed to circulate freely in the origin areas as this would foster indigenous investment or the capitalization and concentration of resources such as land. The money paid to labour can either be returned to the capitalist system through tax or through purchases made in company-controlled stores. Where wages are kept low, leakage from the system will be relatively unimportant, but where these increase through worker pressure, as occurred in South Africa from the late 1960s, the volume of remittances becomes too great to control and the system is undermined. This was surely one factor in the retreat of the labour reserves there, although the situation was compounded by political considerations. With the volume of remittances currently sent back into the labour frontier at such high levels, no centralized control is possible, no matter whether this is attempted by the state or by capital, and the labour reserves are gradually transformed through the entrenchment of a monetized economy. This allows greater independence of movement, providing potential migrants with the means to move in a variety of ways, and the labour reserve becomes the more general labour frontier, which includes the movement of settlers.

Social and political consequences of migration

The impact of remittances is but one consequence of labour migration. There are significant social and political implications too. The separation of families by labour migration with, for example, men in the Middle East and women and children left behind in the villages, can cause problems, although there appears to be no general pattern. There are reports of marital breakdown but also of strengthening of family ties: separation can promote more modern, open attitudes in some, while reinforcing traditional values in others. Much may depend upon the relative success of the migrant and on his or her education and class background. The size of the migrant community at the destination, the cultural difference between migrants and hosts and the institutional framework in which the migrants move are all important factors. Koreans, when South Korea was a significant supplier of labour, tended to move within an enclave community in which everything was supplied by their Korean company. Thus, they worked within a familiar environment but were almost entirely isolated from the host society. Workers in large migrant communities of culturally similar peoples, Muslims from Pakistan working in the Middle East, for example, were likely to face fewer problems than those from cultures with quite different values.

The above issues are given full treatment elsewhere (see Gunatilleke 1986, 1992; also Addleton 1992) but one issue perhaps stands out as more significant for development: changing gender relations. Increasing female participation, in both contract migration and in more independent movement, has been a virtually ubiquitous characteristic of movement out of all parts of the labour frontier. In the cases where men go and women are left behind, the women are often thrust into positions of responsibility, taking decisions independently of their husbands on bringing up children or running the family farm or business. On the return of their husbands after a long period of separation, women may not be so willing to revert to traditional patterns of submission, which may create strains in relationships.

Where women have the chance to migrate, they have opportunities to enter the labour force and gain an independence that would have been inconceivable in their communities of origin as, for example, in Turkey (Gitmez 1991: 131). The migration of women from Mexico to the United States has perhaps benefited women more than men in terms of their spatial mobility, the division of labour, and power and authority, which have all led to more egalitarian gender relations (Hondagneu-Sotelo 1994). Nevertheless, as emphasized earlier, not all migration has necessarily resulted in an improved status for women. The entry into domestic service often leads to the deskilling of female labour and the illegal trafficking in women is simply degradation and exploitation. As a working hypothesis, we can speculate that a greater proportion of the migration resulting in exploitation will come from the more isolated, peripheral parts of the labour frontier, while those movements leading to an improvement in status and an improvement in quality of life more commonly come from those areas closer to expanding core or core areas and from those areas with a longer history of contact with the core.

As money is diffused into the labour frontier in the form of remittances, so too are other, perhaps even more disruptive forces: ideas about different ways of doing things, for example. Circulation and the return migration of people who have spent considerable time away from their home communities characterize the labour frontier. Some 55,000 to 60,000 migrants returned to Turkey every year during the mid-1970s, although their numbers have declined to between 20,000 and 30,000 per annum since then (Gitmez 1991: 123). Over a ten-year period, some 1 million Pakistanis returned (Addleton 1992: 191). We still know very little about the implications for broad social and political change of the continual return of large numbers of migrants over long periods of time. The impact should vary depending upon whether the migrants are moving to core or expanding core destinations, or to areas in the resource niche, as the former are more 'modern' than the latter. The majority of countries in the labour frontier have higher fertility than those in the cores, so do returning migrants introduce new ideas about reproductive behaviour and family size (Díaz-Briquets 1991: 185)? If fertility decline is indeed a result of ideational shift (Cleland and Wilson 1987), then return migrants may be key role players in the dissemination of these new ideas. As stressed earlier, the whole relationship, if any, between patterns of population mobility and changing fertility is poorly understood.

Even less clear is any impact of return migration on the direction of political change. The likely influence of returned students on democratization in new core areas was raised in Chapter 4, but returned migrants, and migrants in general, have often played critical roles in new and often revolutionary movements (Skeldon 1987b). The spread of information about conditions elsewhere can lead to discontent. There is an absolute level of physical deprivation below which human beings cease to function effectively but so much of poverty is the idea of being poor; of relative deprivation and comparing one's own situation with what exists elsewhere. Migration can thus lead to the extension of feelings of being poor, of poverty, when the latter is considered in terms of perception rather than objective criteria of minimum basic needs. The greater awareness of returned migrants, if combined with charismatic personalities and periods of real economic stress, can result in violence, millenarian movements and pressures for change in existing power relations and structures. The returned migrant, schooled in the outside world, can be a force to promote development though remittances, local investment and entrepreneurship, but can also be a force for destruction.

The migrant need not necessarily be a proactive agent of political transformation, however. People may react to changed circumstances by migrating away from exploitation or political unrest. It is one of the 'weapons of the weak' (Scott 1987), whereby the powerless can exert some pressure on the powerful through a withdrawal of their labour and movement out of an area. This thesis of mobility and resistance can be developed at several levels of analysis. For example, in the new cultural geography, Cresswell (1993) argues that Jack Kerouac used mobility as a theme to assail the prevailing American values of the time of rootedness, family values and the 'American dream'. Mobility is perceived as a threat to stability and order and, as emphasized in the Introduction, is seen as abnormal and revolutionary in a way, a threat to the establishment, whether family or government. While mobility might thus be interpreted as a form of resistance to an American ideal of home and rootedness, it must also be stressed that mobility itself is also an American ideal (in contrast to a European ideal), which is encapsulated in the frontier spirit of western expansion, as well as in the visions of the hobo and migrant worker (Allsop 1967). These latter ideals are primarily male values and McDowell (1996), in her critique of Cresswell, is almost certainly correct in allocating gender a more significant role than mobility in any assessment of Kerouac's writings. The debate well illustrates the difficulty of associating migration with any particular set of values.

Given the perceived linkages between migration and resistance and unrest, there is thus a fear among many governments in the labour frontier that large numbers of migrants may return over a short period of time. These fears were realized during and immediately after the Gulf conflict in 1990–1. Some 1,180,000 workers left Kuwait and Iraq during August and September 1990;

540,000 for Egypt and Jordan, 125,000 for India, 60,000 for Bangladesh and 53,000 for Pakistan were among the major flows (United Nations 1996: 209). Not only were the areas of origin of workers suddenly deprived of remittances, but they also had to absorb, in economies with already high unemployment, large numbers of people who had been used to regular employment and high salaries. Social unrest appears to be a possible outcome. So far, however, such fears have largely been unfounded. The majority of those returning appear to have been more concerned with re-migration overseas than with absorption into local labour markets. The observed high rates of unemployment among returned migrants may be more voluntary than forced, as they wait for suitable employment; or they may perhaps not even be actively looking for work at all but enjoying the fruits of their labour overseas (Addleton 1992: 29). Many returnees tend to settle in cities, rather than return to rural home areas, demonstrating a 'J-turn' pattern of migration that will ultimately accentuate urbanization. This pattern reinforces the linkages between internal and international movements, with the international migration merely a spatial extension, a distant node, of the internal rural-to-urban migration fields.

On the interface of the labour frontier: incipient transitions

Proximity to border areas certainly can affect the overall patterns of development positively or negatively. For example, cities in Colombia close to the border with Venezuela experienced a much sharper downturn in their own economies consequent upon the Venezuelan recession than did cities further away, as the turnover in the number of short-term migrants and the amount of remittances declined (Díaz-Briquets 1991: 191–2). As emphasized in Chapter 2, the boundaries between the development tiers are not fixed but can fluctuate back and forth. Parts of the tier defined as the labour frontier could perhaps be considered part of a potential core and other areas part of the resource niche. Turkey could well be placed in the former category. As seen earlier, it has played an important role in the migration of labourers to Europe, primarily Germany, and also to the Middle East (Martin 1991, Gitmez 1991).

One of the basic premises underlying this book is that there cannot be migration without some kind of development. Hence, a statement such as 'migration without development: the case of Turkey' (Gitmez 1991) must either be exceptional or flawed. The thrust of the argument behind that statement was that the remittances from Turkish workers were declining as more and more families took up long-term residence in Europe. The decline in remittances had to continue as fewer and fewer opportunities existed overseas for future labour migration. Moreover, the remittances that had been sent back had not been used for the most productive of purposes. The labour migration had not only made little impact on the overall rate of unemployment but might have added further to it as greater numbers of migrants returned to Turkey. Even the data presented by Gitmez (1991: 129), however, do not support such a pessimistic scenario as he argued that the country had been integrated into the world economy during the period of labour migration. Also, the Turkish economy, as measured in growth of GNP per capita, had been growing since 1980 at a faster pace than most of the European economies, and faster than most of the non-East Asian economies ranked above it on the World Bank table (World Bank 1995).

Turkey has itself become a target of considerable migration, particularly since the opening-up of eastern Europe and the ex-Soviet Union since 1989. To be sure, some of this migration is an acceleration of long-standing return movement of ethnic Turks, from Bulgaria in particular, which saw a quarter of a million return in the late nineteenth and early twentieth centuries, and continued with marked fluctuations over subsequent decades to reach hundreds of thousands again in a wave of migration in 1989–90 (SOPEMI 1994: 111–14). Perhaps 300,000 entered Turkey, with about 100,000 subsequently returning to Bulgaria. The majority were highly educated, with large numbers of professionals attracted to Turkey, 'a prosperous state with a developed market economy and relatively great demand for highly qualified and skilled labour' (SOPEMI 1994: 113). In addition, hundreds of thousands of former Soviet citizens came, mainly from Georgia but also from other republics, to 'sojourn' in eastern Turkey. While many are seeking entry into Turkey's labour markets, large numbers are self-employed in the form of petty traders who virtually commute into the country and whose average stay appears to be fifteen to twenty days (Aktar and Ogelman 1994: 346). Thus, Turkey appears to be on its way to a transition from emigration to immigration, as it becomes integrated into regional and global economies, although the high incidence of short-term circulation into the country and persistent outmigration suggests that this is as yet at an early phase. Whether Turkey indeed achieves the transition will depend as much on the strength of its recent democratic system in the face of Islamic challenges and minority group pressures as on continued economic growth.

Other parts of the labour frontier that may see rapid future development that could give rise to an eventual turnaround in migration and their inclusion in a future expanding core are Indonesia and the Philippines. However, studies of labour markets in these nations suggest that both are some considerable way from any transition towards a slowing in emigration or any trend towards immigration (Manning 1995, Amjad 1996). Both are likely to remain areas of labour surplus and will continue to supply labour to the rapidly expanding cores overseas, as well as to the local urban complexes around Jakarta and Manila.

On the interface of the labour frontier: reversing transitions

Much of the increasing migration to Turkey observed above has been the direct result of the structural changes that have taken place after the opening-up of eastern Europe in 1989. Of all the parts of the labour frontier, only the countries of eastern Europe and the southern and western parts of the ex-Soviet Union have had persistent low fertility and low population growth. Their demographic profile is not fundamentally different from that of the countries of western Europe and so the recent movement out of these areas is certainly not propelled by excessive population growth. On the contrary, eastern European countries resorted to importing labour to offset low growth in their labour forces and to help industrial development. The labour came mainly from Cuba and Viet Nam, and to a lesser extent from Nicaragua and Ethiopia (Pérez-López and Díaz-Briquets 1990). As these cases and the example of China discussed earlier in the chapter illustrate, labour migration was not a characteristic solely of capitalist economies.

Before the separation of the Second World from the First World at the end of the Second World War, the countries of central and eastern Europe had been a traditional source of manpower for the industrial countries of the west (Grecic 1993: 139). Eastern Europe was a labour frontier for the west. The Second World War and its immediate aftermath saw massive movement of peoples, much of it along ethnic lines (Clout and Salt 1976), and the haemorrhaging of refugee populations continued until 1961 with the construction of the Berlin Wall and the fence between the then Federal Republic of Germany and the German Democratic Republic (GDR). Before then, some 3.7 million Germans, about one fifth of the population of the GDR, crossed over to the west, many of them highly educated and skilled (Rhode 1993), and 200,000 fled Hungary after the 1956 uprising. Net emigration was an important factor in the decline in the GDR's population in the 1960s, the only country in eastern Europe where this was so

(Compton 1976). Movement out of the region, except for Poland, was tightly controlled until the late 1980s, since when the migration has recommenced sharply.

Internal movements within eastern Europe, unlike those in the Soviet Union, declined sharply from the mid-1950s (Fuchs and Demko 1978: 171). The principal reason was the substitution of commuting for migration. While this fits the general idea of a mobility transition, it occurred at much lower levels of urbanization than in the west because of socialist government policies affecting the location of industry and housing (Fuchs and Demko 1978). It was extremely difficult to move into cities because of the lack of housing. The renewed emigration, from Poland in particular, was strongly urban in origin (Korcelli 1992: 302). Whether this will open up a housing market to transform commuting to urbanward migration remains to be seen. To add to the complexity of the flows in this part of the labour frontier, there are movements from further east into Poland and Hungary in a vast system of westward stage migration.

In Chapter 5, parts of western Russia and of eastern Europe were designated as a restructuring core. The extent to which these areas will ever realize this status will depend to a large extent on their ability to retain their labour. At present, most of the region can be seen to have returned to its traditional role as a supplier of labour, but a supplier to economies which are labour-surplus rather than labour-deficit. The migration from Poland over the 1980s was so intense that it exceeded natural increase, causing a decline in the population (Grecic 1993: 146), although there are signs that the emigration is waning in the 1990s, with immigration and return migration increasing (SOPEMI 1993: 134–5). Real growth in GDP in 1994 was the highest for several years and, with a low-cost, skilled labour force, eastern Europe should be able to mirror, albeit palely, the successes of East Asia. In the past, eastern Europe and much of western Russia moved from labour frontier to a core of a separate socialist system and back to a labour frontier as the areas became incorporated into a global system. Whether they can achieve sustained faster growth and a transition from emigration to immigration will depend primarily upon the creation of a favourable investment climate for international capital.

On the interface of the labour frontier: the agricultural frontier

At the other end of the spectrum, there is migration from the labour frontier into the resource niche. This will be discussed much more fully in Chapter 7, in the context of the migration to the Middle East, but the critical difference compared with the movement to the expanding core and core areas is that there is virtually no scope for more permanent migration. All workers must return to their home countries, precluding the development of permanent linkages and diasporic communities between these development tiers. The only cases of permanent settlement are in the context of internal movements to the agricultural frontier, most notably in the transmigration projects in Indonesia and avenues of penetration into the Amazon in South America. Often used as a safety valve to settle landless peasants from densely populated areas, these projects are perhaps better conceptualized as extensions of national sovereignty whereby nations attempt to occupy and populate fully the territories to which they lay claim. The numbers moved can be very substantial.

The transmigration programmes in Indonesia have been among the largest attempts at resettlement in the world. Dating back to 1905, the Dutch colonial authorities embarked upon projects to move people away from the densely populated islands of Java and Bali towards the sparsely settled lands in Sumatra, Kalimantan (Borneo) and Sulawesi. About 100,000 people had been resettled by 1930. The programmes accelerated after independence, and particularly after President Suharto adopted transmigration as one of his personal projects (Tirtosudarmo 1990). Perhaps 193,700 families were resettled between 1950 and 1979, when there was a surge in the programmes and 366,000 families, involving some 1.5 million people, were moved between 1979 and 1984. These figures refer to sponsored moves. Perhaps an equivalent number of others moved to the outer islands as spontaneous migrants. An additional 179,000 sponsored and almost 300,000 spontaneous and partially assisted families moved in the three-year period to 1987, after which the programmes were scaled back in the face of increasing evidence of environmental damage (Bilsborrow 1992).

While the Indonesian programme is perhaps exceptional in its magnitude, it is typical in its intent, as many countries have pushed their agricultural frontiers into sparsely settled and remote areas. The only other scheme comparable in scale was the sponsored westward movement of Chinese settlers into Xinjiang, where there was a net inmigration of about 3 million people during the communist period to 1982 (Banister 1987: 311). The success of these schemes has been variable and must be measured along several dimensions: resolving overpopulation and landlessness in areas of origin; diverting migrants away from the largest cities; increasing agricultural production; establishing stable, thriving communities in areas of destination; and extending the area under effective central government control. The last of these is the most often attained, the first two the most rarely achieved, and the success of the other two objectives depends very much on the amount of investment in infrastructure and support services that a government is willing to make.

One of the most successful attempts was the FELDA land settlement scheme in Malaysia, which met most of the above objectives, but at a cost of between $US6000 and $US9000 per settler (Baydar et al 1990). Few countries have financial resources of this magnitude to invest in land settlement or the organizational ability to implement such schemes. China, with its strong central control during the Maoist period, was also able to open up new lands through the westward redistribution of its population at that time. 'In essence, the contribution of the directed interprovincial migrants to China's development has been to diffuse modern techniques and ideas' (Banister 1987: 312). Elsewhere, and more generally, the schemes are established on marginal tropical lands; failure rates and return migration to the home communities are high. Spontaneous settlements often have a higher probability of success than those planned by government, perhaps because of the higher motivation of independent settlers. High fertility on the agricultural frontier often means that local resources cannot support the next generation, who join the migrants to the urban sector.

The magnitude and widespread distribution of land settlement schemes within the labour frontier draw attention to the importance of migration within the rural sector. Most of this is not movement to agricultural frontiers but the movement of labour to plantations or other centres of commercial agriculture. Most of all it is the constant interchange of population for marriage and family formation and dissolution that makes up the normal cycle of life for millions of peasants throughout the labour frontier. These movements will become still more pronounced as we move on to consider the most rural of the development tiers, the resource niche.

Chapter 7

The resource niche

Within all the development tiers discussed so far there has been considerable variation, although all have had a functional unity of a sort. It might be thought that this last development tier, the resource niche, is no more than a convenient catch-all in which to consider the remaining parts of the world. What could the highly urbanized desert kingdoms of the Middle East, the dispersed hunter-gatherers of the Amazonian rainforests and the populations of the islands of the Pacific or the Caribbean possibly have in common? There are obviously huge differences in terms of cultures and development within this tier, between the futuristic architecture of some of the Gulf states, the luxury tourist resorts of Polynesia and the sun-dried clay houses of the Sahel. It would indeed be possible to create a whole series of developmental regions within this tier, but several common threads do seem to run through the diversity. The first is vulnerability: a vulnerability in terms of environmental issues, in terms of size and in terms of a tenuous link with the other development tiers, and particularly the old and new cores. Although, as we shall see, the volume of migration may not always rival that in the other tiers, it is in this development tier that the relative impact of the movement is perhaps the greatest. The most vivid common thread is that this tier has been and will be, in the foreseeable future, little more than a resource niche for groups in the first three development tiers.

As a resource niche, its population is also clearly a resource, perhaps the 'ultimate resource' (Simon 1981), and there should therefore be clear similarities with the labour frontier, just discussed. In some areas, this is indeed the case, but generally the populations are too small, too dispersed and too isolated from centres of economic growth to make them an important resource at present. The exploitation of physical resources, renewable and non-renewable, for the benefit of core areas and expanding cores remains the dominant feature of this tier.

This development tier includes some of the poorest economies in the world and while the physical environment and disease, to take but two groups of factors, obviously have their impact in other development tiers, the lack of financial resources available to deal with them means that their consequences in this development tier are more acute. Perhaps more critically, the structure of the state in this tier either tends to be weak, in some areas virtually nonexistent, or under threat from competing interests. As a result of environmental deterioration and political instability, most of the world's forced population movements, both the internally displaced and refugees, are either found in this tier, or created by conditions internal to it. The political and physical environments are much more fragile than in the other development tiers.

There are large urban centres within the tier but, with the exception of the Gulf states and some of the small island countries, these regions are still primarily rural. It is in this tier, and particularly in sub-Saharan Africa, that the distinction between internal and international migration becomes most blurred. The national boundaries established by the colonial powers and the basis of the independent nation states of today bisected ethnic groups and traditional trading and grazing circuits. The Fulani of West Africa now live in eight countries, and single ethnic group nations are unknown (see Adepoju 1991). The present patterns of migration are exceedingly complex: much of the movement is circular, of varying duration, and much is within rural areas, although there is urbanward movement to local centres as well as out of the tier. However, migration away from most parts of the tier is relatively unimportant compared with movement within it, and many of these areas are only tenuously linked to the global system. Sub-Saharan Africa's share of world trade and investment, for example, has 'declined to negligible proportions' (Collier 1995: 541). Although the resource niche has the above common features, its diversity prompts a subdivision into three general areas for discussion: regions of deterioration covering much of sub-Saharan Africa; small regions covering islands in the Pacific and the Caribbean, as well as isolated mountain and continental communities in Central Asia, for example; and the oil-rich communities of West Asia. All, however, are regions at risk.

Regions of deterioration: reversals in development

Sub-Saharan Africa not only includes some of the poorest countries in the world but has also seen a steady decline in its standard of living over the last thirty years. In 1965, 'Ghanaians were less poor than South Koreans and Thais and Nigerians were better off than Indonesians' (Adepoju 1995a: 324). This statement tells us as much about the tremendous progress made in East Asia as it does about the situation in sub-Saharan Africa but, for many Africans, living conditions in the 1990s appear to be worse than at the time of independence in the 1960s. Persistently high population growth, deteriorating terms of trade, lack of reform, and corrupt and ineffectual governments have all contributed to a deterioration in levels of economic development. The situation has certainly not been helped by environmental conditions, desertification and failure of the rains, but these physical difficulties themselves have been exacerbated by rising populations and overgrazing, and by warfare, which brought severe famine in its wake in several parts of the region.

Given the negative trend to development, there may be significant differences from the patterns of population mobility previously described for the core areas. The major impression from the migration literature is that sub-Saharan Africa is an area of tremendous mobility. With a population representing about 10 per cent of the world's total, the region has been considered to account for as much as 40 per cent of the world's migrants (Russell, Jacobsen and Stanley 1990: 14). A total of 35 million international migrants for the late 1980s, first estimated by Ricca (1989), is found 'plausible' by these latter authorities (see also Stalker 1994: 233), although judged high by Tapinos (cited in Coussy 1994: 240). The methods of calculating the numbers from the very inadequate data available are reasonable but should not be used to imply that migration in sub-Saharan Africa is any more, or any less for that matter, important than anywhere else. We have already seen the high levels of mobility in pre-modern Europe and Asia (Chapters 3 and 4) and although Africa has been, and is, a continent on the move, whether rates of mobility there are any higher or lower than anywhere else is impossible to say from existing data. The large proportion of total international migration accounted for by the sub-Saharan numbers is due primarily to the widespread bisection of traditional identity groups by the recent, in historical terms, imposition of national boundaries. What the data do reinforce is that high mobility is not caused by modern development and that migration has been an integral and everyday part of the life of all societies.

Of much greater importance than looking at the absolute numbers of migrants is to consider the patterns of movement. Excluding the countries of North Africa and South Africa (considered as part of the labour frontier in Chapter 6), Africans represent only a minuscule proportion of the foreign-born in the major destination countries: just over 1 per cent in the United States in 1990, 2 per cent in Canada in 1991 and 0.7 per cent in Australia in 1991, many of whom were whites from Zimbabwe. Because of its historical involvement in Africa, France is the major destination of movement out of Africa, but even there sub-Saharan Africans represented only about 4 per cent of the foreign-born in 1982 (Garson 1992), although they were increasing rapidly and accounted for over 10 per cent of the annual inflow in the 1990s (Tribalat 1992, 1996). Compared with the countries of the labour frontier discussed in

Chapter 6, the countries of sub-Saharan Africa have been, and generally still are, quite weakly integrated into the global migration system. Most of the movement is internal to the region.

As previously emphasized, much of the migration is along traditional nomadic or seasonal paths, particularly in West Africa, where the north–south ecological zonation of desert, grasslands and forest channelled movement in the search for fresh pasture or in the exchange of products. Of course, not all of the migration captured by the data refers to traditional forms of mobility. As in Asia and Latin America, the introduction of commercial agriculture, in the form of plantations, and the exploitation of minerals introduced new types of migration, chiefly labour migration of the type described in Chapter 6. The resultant movements were primarily north to south, to the centres established by Europeans along the coast, migrations that established a symbiotic relationship between two ecological zones. The period of slack labour demand in the agricultural cycle of the dry north coincided with the period of peak demand in the wet south.

Assessments of migration and development in West Africa made from the early 1960s to the mid-1970s would have seen the makings of an integrated developmental region with a labour frontier in the northern parts of the region and significant poles of attraction, even potential expanding cores, initially in Ghana, the Ivory Coast and, to a lesser extent, Senegal and, in the 1970s, the oil boom in Nigeria. Many of these flows are described in Zachariah and Conde (1981) and they led to a change in the balance of population between the interior and the coast from 50/50 to 33/66 in the fifty years from 1920. 'There is no other known region of this size in the world that has in one half century undergone such a considerable transfer of population' (Amin 1995: 36). Initially consisting mainly of the circulation of males to plantations, especially in the cocoa belt, the movements became more long-distance and more long-term, and included women and children, in a migration to the cities. The source areas lost their most dynamic and youthful populations while the coastal areas gained (Amin 1995). These movements were both internal and international, but in Ghana in 1960 there were 828,000 foreign nationals, equivalent to 12.3 per cent of the national population, the majority from Upper Volta but also from northern Togo. The Ivory Coast became the principal destination from the 1960s and by 1975 over 20 per cent of its population was immigrant with the largest numbers from Upper Volta, Mali and Guinea (Zachariah and Conde 1981: 35). In the early 1980s, there were some 2.5 million migrants in Nigeria, about 2.5 per cent of the total population, although these estimates were probably low (Russell, Jacobsen and Stanley 1990, vol II: 62). The tradition of a free flow of labour was recognized in the foundation of the Economic Community of West African States, ECOWAS, in

1975, although its ideals were shortly to founder on the reality of economic crisis.

The free flow of labour between the colonies gave way to more restrictive policies upon the assertive nationalism that came to typify West African countries after independence. The reasons for the implementation of policies to remove non-nationals were usually couched in economic terms, but they were introduced to protect or improve the position of the dominant national ethnic group. For example, the fact that there were some 600,000 registered unemployed in Ghana in the late 1960s was used as a justification to reduce the number through emigration. Also, the money remitted to their home countries by migrants was seen to aggravate the balance-of-payments deficit at that time (Adomako-Sarfoh 1974: 139). Within about six months of the announcement, in November 1969, that non-nationals without valid residence permits would have to leave, some 213,750 had left Ghana. In the decade of the 1960s, Ghana is estimated to have had a net emigration of 400,000 (Zachariah and Conde 1981: 45). The migrants, however, were to be found in virtually every sector of Ghana's economy but particularly in the staple cocoa industry and in retailing and distribution. Their expulsion certainly had adverse short-term effects on the cocoa industry (Adomako-Sarfoh 1974) but more long-term repercussions on the economy as a whole. Between 1960 and 1982, Ghana recorded a growth in per capita GNP of −1.3 per cent per annum, which contrasted sharply with the positive growth of around 2 per cent per annum in neighbouring economies at this time (Adepoju 1995a: 348). As a result, much of Ghana's own skilled labour left and by 1981 '9 per cent of Ghanians and about 25 per cent of its labour force had emigrated' (Adepoju 1995a: 348).

The booming Nigerian economy absorbed many of the skilled workers leaving Ghana but, by 1983, the boom was over and that economy, too, was near collapse. Again, the immigrants were targeted, with between 1.2 million and 2 million being expelled in 1983, and skilled Nigerians began to leave the country in increasing numbers. Hence, in parts of West Africa, the general trend from emigration towards immigration observed in the core and expanding core areas does appear to have had its mirror image where countries which were once prosperous and points of attraction for immigration then became countries of emigration. This reverse migration transition applies particularly to Nigeria and Ghana, which were among the principal poles of attraction. Côte d'Ivoire, the other main destination pole in west Africa, with fully 26 per cent of its population foreign-born in 1988, may have seen a slowing in its immigration since then. There was a 'massive' outmigration following the devaluation of the currency in 1994 (Makinwa-Adebusoye 1995: 447) but, as conditions in neighbouring countries are worse, there are still strong pressures to move to that centre of relative prosperity. The government of Côte d'Ivoire is one that considers its immigration 'too high' (Russell, Jacobsen and Stanley 1990) and thus at least a partial reversal may be seen in that country too.

Despite the lack of a reliable database, there are indications that the reversal is much more widespread if we consider the pattern of urban growth. Potts (1995), in a wide-ranging review was urban growth in sub-Saharan Africa as a whole in the 1980s, argues that the contraction of the urban–rural wage difference not only produced a return movement to the rural sector but discouraged people from moving to towns in the first place. Thus, cities that were once dynamic centres of growth (and for a classic study of urbanization in Ghana under these conditions, see Caldwell 1969) had become centres of stagnation and sources of migration. This reversal of rural-to-urban movement reinforces the disarticulation of the rural sector in sub-Saharan Africa from more developed centres discussed in Chapter 6, where we saw the retreat of the labour frontier in southern Africa back to concentrate on Lesotho, Swaziland, Botswana and the very southernmost parts of Mozambique. The rest of Mozambique, Angola, Malawi and Zimbabwe became separated from the system.

The return migration was not simply of 'the poor and unemployed who are leaving town' (Potts 1995: 259) and going back to the villages, but also the highly educated who were leaving, or attempting to leave, these countries. The brain drain from sub-Saharan Africa, like that from eastern Europe, indeed may have strong negative consequences for the countries concerned. This situation is quite different from that in Asia, where the movement of the skilled and educated, as seen in Chapter 4, was definitely positive for development. Up to 1987, perhaps 70,000 high-level emigrants had left sub-Saharan Africa, small in terms of the total population of the region, but significant relative to the pool of skilled labour available representing '30 per cent of the highly skilled manpower stock' (Adepoju 1995b: 99). The majority appear to have left for the European Community, following old colonial ties to Britain and France. While it would be wrong to attribute the deterioration in conditions to emigration alone, there seems to be little question that it has exacerbated the economic situation and that these reinforce each other in a negative development spiral. People leave because of deteriorating economic conditions, while these become even worse because those with ability and entrepreneurial skills are leaving.

While the economic conditions have been emphasized, these must be viewed against the backdrop of ethnic tensions within the context of newly independent nations, as suggested above for the case of the expulsions from Ghana and Nigeria. There, action was taken against immigrant groups: a growing tendency is for the tensions to be generated between groups native to nations themselves. It is in East Africa that these tensions seem to have made their greatest impact to date although, unfortunately, tensions do appear to be growing in West Africa too, as the recent conflict in Liberia and the repression in the largest country in sub-Saharan Africa, Nigeria, so clearly show. These tensions are not new in any way, as the Biafran War in Nigeria in 1967 demonstrated but, until the late 1980s, the state appeared to hold as an integral unit in the region. Now that ability is under severe question.

Regions of deterioration: refugees and displacement

East Africa, more than any other region, is associated with refugees, generating 72.6 per cent of the world's total in 1991 (Russell cited in Oucho 1995: 396), while accounting for only about 4 per cent of the world's population. In part, this high proportion is deceptive, as there are clearly people in many other parts of the world who have been forced from their homes but who do not appear in the statistics on refugees. For example, the internally displaced in large nations such as China, Mexico or Indonesia do not appear in the figures, irrespective of whether they have moved for political or environmental reasons. East Africa, with a number of small states, is going to generate a higher number of refugees simply because the displaced have a higher probability of reaching and crossing a national frontier, and hence of achieving refugee status. Despite these purely methodological issues, there are three factors that characterize East Africa which provide fertile ground for the creation of refugees. First is the patchwork of competing ethnic groups that makes up the region. Second is the disintegration of the state as an effective instrument of government in several parts of the region, Somalia and Rwanda being the most obvious examples. Third is the existence of repressive regimes in other parts of the region, often ruled by representatives of minority groups.

Civil war and inter-ethnic conflict are the culmination of the above factors and the leading cause of refugee flows: from Zimbabwe in 1978–9, Ethiopia in 1987–9, Mozambique also in 1987–9, Uganda in 1978–9 and 1989, Sudan in 1986–9 and Rwanda in 1978–9, 1987–8 and 1994 to the present (Oucho 1995: 397). One of the earlier significant flows was of the East African Asians, mainly from Uganda but also from Kenya and Tanzania between 1969 and 1972. This refugee flow was significant from two points of view. First, refugees were expelled mainly to core countries, and also to India, rather than across the border to neighbouring countries, as in the case of most subsequent refugee movements. Second, this was an expulsion of relatively wealthy, professional and entrepreneurial groups, whose leaving had a profound impact on already weakened economies and accelerated the deterioration (Zolberg, Suhrke and Aguayo 1989: 65–6; Robinson 1995). More than 80,000 Asians in Uganda appear to have left over the three-year period and only between 1000 and 2000 remained after the 1972 expulsion.

Refugees are obviously found elsewhere in the world – from Afghanistan, Tibet and the mountainous regions of the Caucasus, for example – but again they tend to be concentrated in this resource niche development tier. The only major centres of refugees outside this development tier and on the margins of a core area are those from the republics of the former Yugoslavia, where again ethnic hatred, which goes back centuries, has resurfaced upon the collapse of central government control, and around the borders of Israel, where over 1.3 million Palestinians are still to be found (Adelman 1995: 415). The intense diplomatic efforts to end these disruptions close to core or potential core areas are in stark contrast to the indifference with which the international community, apart from humanitarian relief agencies, has viewed so much of the conflict in sub-Saharan Africa. The latter's position, far from the main centres of economic dynamism, particularly now that the rivalry between First and Second worlds in the periphery is over, has rendered this area of little importance internationally, save as a future resource niche.

Development and refugees are mutually exclusive. Where development has been most rapid, as in the rapidly expanding core areas of East Asia discussed in Chapters 4 and 5, the refugee, excluding the internally displaced, has become virtually a person of the past, although certainly only the recent past. The expansion of capital requires stability and the creation of an environment where risk can be managed. That expansion will either avoid, or seek to alter, areas where conditions exist that are likely to generate unrest on a large scale, and thus refugees. With the international community favouring the repatriation of refugees as the most desirable solution rather than the granting of asylum in stable areas, the problem of refugees is being increasingly contained within the poorest areas themselves. The issue of refugees in a world of increasingly unsympathetic borders is considered fully in Richmond (1994) and Zolberg, Suhrke and Aguayo (1989).

Development itself can, however, create forced movements, although these are more likely to be the internally displaced rather than refugees. For example, the huge Three Gorges scheme to dam the Yangtze in central China will displace between 700,000 and 1.2

million people (Douglas, Gu and He 1994: 194). Also in China, it has been estimated that the generation of every 100 GWh of electricity from large and medium-sized power plants required the flooding of 50 hectares of farmland and the resettlement of 560 people which, given the recent history of dam construction, would imply the displacement of 450,000 people (see Smil 1992: 14). Massive though some of these schemes deliberately modifying the environment might be, their impact on population movements is small compared with the more long-term, often gradual, alteration of the environment by human and natural agency. Again, these impacts are most acutely felt in this development tier. The desertification around the margins of the Sahel or in western China, the destruction of the tropical moist forest in Amazonia or central Africa and the devastation of hill lands in Nepal and northern Thailand are some of the more significant examples. Except in extreme cases where conditions have given rise to famine – and even here famines can seldom be attributed solely to strictly environmental causes – it is difficult to explain migration by purely physical factors, save in the case of earthquakes, volcanic eruptions, cyclones or tsunamis.

Climatic conditions in the Sahel have moved populations southwards to areas with more favourable rainfall (Findley et al 1995) and the periodic droughts have pushed people into towns (Thiam 1994). These issues, however, revolve as much around increasing populations, bringing pressure on scarce resources, and lack of investment as they do on physical geography. Where central authority extends into these marginal lands, as in western China, programmes of resettlement from the degraded lands can be attempted. One of the largest involved the resettlement of 320,000 people from the northwest to irrigated lands in central China in 1990, a programme that was likely to cause resentment among the populations in the destination areas (Smil 1992: 14). The environmental impact on migration is well reviewed in Suhrke (1993). Refugee movements can have a devastating impact on the environment. For example, for every 100,000 arrivals, it is estimated that 333 hectares of land needs to be cleared and the refugees will require 85,000 metric tons of fuelwood every year (Ferris 1993: 104).

Although forced migrations are a major part of total mobility in this development tier, they are not the only or necessarily the most important type of migration. Non-permanent and circular forms of movement do tend to dominate and many of the issues discussed in Chapter 6 also apply in this tier. Remittances are one of the 'most pervasive' aspects of the African migration systems and are characterized as 'fundamentally a family and not an individual activity' (Adepoju 1995b: 100). As in the labour frontier, remittances play an important role in supporting households in the areas of origin, although the migration in this tier appears to be much more a survival strategy than a means of improvement. Unlike the labour frontier, remittances play an important role at the macrolevel only in a few countries such as Burkina Faso, Mali and Benin, where they make a major contribution to foreign exchange earnings. The sources of the remittances are primarily neighbouring countries, rather than distant or core areas as in the labour frontier.

In common with migration in other tiers, there has been an increasing participation of women in the movements. Migration in sub-Saharan Africa was for long the preserve of males, with women playing the more important role in the domestic agricultural economy. Throughout this development tier, and much more pronounced than in the other tiers, is a sharp division of labour by sex. Plantation agriculture and mining demanded male labour, with women heading the households back home. Although data are scarce, the indications are that there has been increasing movement of women independent of, or even despite, the wishes of their menfolk, which has been facilitated by these separate roles: women can only fulfil their particular obligations through migration (Findley et al 1995: 489–90). They are not, and have not been for some considerable time, passive victims of either traditional gender relations or of the new economic system introduced in the colonial and post-colonial periods (Nelson 1992, Brydon 1992). While there may appear to be similarities with the situation in the labour frontier described in Chapter 6, these movements appear to be more the result of deteriorating conditions in the rural areas than of increasing opportunities opening up in potential destination areas. The domestic economy through which women traditionally made their contribution has become so undermined by the various crises that they have to seek new avenues through migration. Presumably, though, women are also to be found in the return flows from towns described by Potts (1995) as the result of the latest economic crisis.

Regions of deterioration: the spread of disease

Migration has generally been seen as positive for development or, in the case of this development tier, as essential for survival through the sending of remittances and the diffusion of ideas that can promote improvements in areas of origin. However, not all the consequences of the linkages created through migration need necessarily be seen as benign. In Chapter 4, the potential for migrants to act as agents of violent political change was raised, but there are other potentially disruptive consequences of migration. One of the major issues that will confront the human population as we move into the twenty-first century is the diffusion of certain types of virulent diseases. The massive interchange of population, estimated at perhaps a million person-moves a week between the industrial and the less developed world alone, facilitates the diffusion of disease-bearing organisms into populations which have no natural resistance to their effects (Garrett 1996). The huge cities, in which a majority of the human population will shortly live, provide ideal environments for the rapid spread of disease.

The issue of migration and disease is obviously not restricted to the resource niche development tier but is global in implication. Cases of western travellers contracting exotic diseases in isolated parts of the world and returning home to die in an advanced country capture public attention and provoke fear. The impact that the diseases have is, however, much greater where the provision of health services and standards of public hygiene are weakly developed and where the nutritional status of the population is low. Although great progress has been made throughout the world in disease control, there are still significant spatial differences in the diffusion of the epidemiological transition (Omran 1971), with infectious disease still a leading cause of death at the bottom of the development spectrum and the prevalence of degenerative diseases at the top. The circulation of people from areas of infection is going to be much more effective as a medium for the spread of disease within the resource niche than between the niche and other development tiers. As we have already seen, movements out of this tier are already relatively small and their impact should, theoretically, be easier to control.

Migration leading to the exposure of populations to diseases against which they have no or little immunity is not new. The spread of ideas along trading routes has always been accompanied by the spread of disease (Curtin 1989). Smallpox brought by the early European conquerors of the Americas was one of the most significant contributors to the destruction of the civilizations there on a scale that has no equivalent today. The populations of the Americas were reduced by up to 90 per cent during the 150 years following the Conquest (evidence reviewed in Denevan 1992). Perhaps closest in modern times would be the influenza pandemic of 1918–19, which led directly to the deaths of at least 20 million people, although some argue that this number died in India alone from the epidemic (see Noble 1982: 15). The total may have been closer to 40 million (Shortridge 1995: 1210). Its origins remain obscure, whether in the United States, West Africa, or in a series of places virtually simultaneously, but its spread was unquestionably favoured by the large number of soldiers moving around a world weakened by four years of warfare. The pandemic of Asian influenza of 1957 was more closely monitored; its diffusion out of southern China to Hong

Kong and from there around the world by ships' passengers and crew to other ports, and thence inland, provides a clear relationship between migration and the spread of disease (Fig. 7.1).

At present, the fear is either that some exotic disease such as Ebola or Lassa fever will be brought from the periphery to wreak havoc in the cities of the core or that there will be the rapid spread across the globe of a new virulent type of a common disease such as influenza (Shortridge 1995). When the 'old' disease of pneumonic plague resurfaced in Surat in western India in 1994, some half a million people left the city within forty-eight hours to every part of the country (Garrett 1996: 73). If that disease had been resistant to standard forms of treatment, there might indeed have been the makings of a major catastrophe.

Without in any way denying the very real impact that migration can have on the spread of disease, it is only too easy to exaggerate these dangers in the case of such an emotive disease as AIDS. 'We have not seen any significant epidemic [of AIDS] triggered by the arrival of immigrants from another country', although within areas of high prevalence, 'there is a strong correlation between HIV infection and migration status' (Decosas et al 1995: 826). The researchers found that HIV prevalence among those in a rural community in Uganda who had never moved was around 5.5 per cent, compared with 16.3 per cent among those coming into the community from outside. The highest incidences of HIV prevalence are recorded in a broad swathe of sub-Saharan Africa, with the peak rates in eastern and central parts (Barnett and Blaikie 1992: 23–4). The separate systems of gender circulation, described above, push both male and female migrants into high-risk behaviour. Concentrations of young single men on plantations and in towns, and a lack of opportunities for young, poorly educated women in the rural sector, generate the demand for, and supply of, prostitutes: the ideal medium for the spread of HIV infection.

The return of male and female migrants to their home villages further spreads the infection as does the well-publicized diffusion by truck-drivers along the main roads of a country (Barnett and Blaikie 1992: 28). The impact can be profound, with whole families losing their economically active population and cultivated areas contracting in once prosperous regions (see Gould 1995: 136). While the impact in rural parts of sub-Saharan Africa can be dire and the greatest absolute number of carriers is to be found in that sector, the greatest impact, as far as development is concerned, is perhaps on the educated urban elite. That elite is a relatively small component of the total labour force. As seen above, it has already been depleted through migration and it is being further reduced through disease. HIV prevalence rates are considerably higher among the better educated and wealthier groups,

Fig. 7.1 The diffusion of the Asian flu pandemic, 1957

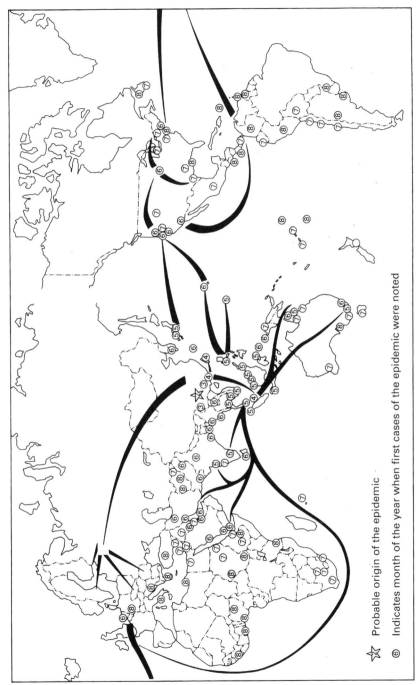

☆ Probable origin of the epidemic

⑥ Indicates month of the year when first cases of the epidemic were noted

Source: The UNESCO Courier, May 1958.

particularly males, in sub-Saharan African countries and the cost of their medical care, not to mention lost production, will be a major factor in reducing per capita income in these areas (Ainsworth and Over 1994). Throughout sub-Saharan Africa, the rising death rates due to AIDS will reduce the rate of growth of the total populations. United Nations estimates (1996) suggest that, by the year 2005, these populations will number about 12 million, or 3.8 per cent fewer than if there had been no AIDS deaths. The increased mortality is also likely to be reflected in a slowing in the rate of growth of those migrating and this, together with a fear of catching the disease in cities, may contribute further to the slowing of urbanization observed earlier.

Clearly, disease does not respect boundaries, and these issues afflict the labour frontier and core areas too. Yet, in these areas, and particularly the cores, the populations are larger, the levels of education higher and the available financial resources greater, so the impact of the disease can be controlled more effectively. Bangkok, in the centre of a rapidly expanding core area, is renowned for its flourishing sex industry. The sex ratios of the 565,261 Japanese and 308,305 German visitors to Thailand in 1993 were respectively 203 and 169 males per 100 females. As yet, there have been no major epidemics of HIV infection in Japan or Germany that can be attributed to these visits. Nevertheless, the infection can be spread into core areas: for example, the original spread of the disease out of Africa was almost certainly caused by Haitian professionals, and the disease later spread from Haiti to the United States.

What is more important, from an epidemiological point of view, is not the spread of the disease but the behaviour of those who become infected. Rarely are the conditions conducive for the explosive spread of HIV between widely separated origin and destination areas. Granted, this scenario has existed, as in the early spread within homosexual communities, but the sex tours to Thailand are unlikely to give rise to widespread epidemics in core areas, because the cultural practices in those areas are unlikely to facilitate its spread. Prostitution involving foreigners represents only a small fraction of total prostitution in the country (Weniger et al 1991: 574) and its most rapid spread is in the poorer peripheral areas. The practices that facilitate the spread of AIDS owe as much to lack of development as to tradition and choice. Prostitution is one of the few means by which women in deprived and impoverished positions can earn enough to support their families. The HIV problem in Thailand will thus remain concentrated there, and primarily among the poor, uneducated women mainly from the peripheral north and northeast, whose infection rates are highest and who are patronized by low-income, often migrant, Thai men in regional cities, as well as in Bangkok. As in Africa, these women become circular migrants because there are no opportunities for them in their villages (Archavanitkul and Guest 1994).

While it is the more dramatic or exotic diseases such as AIDS, Ebola and plague that have captured so much attention, the impact of migration on the spread of infection is likely to be even greater in those areas where long-established diseases such as malaria, trypanosomiasis, cholera and schistosomiasis are endemic. There are currently somewhere between 300 million and 500 million cases of malaria each year, which lead to between 1.5 million and 3 million deaths per annum, for example (Prothero 1995), compared with an estimated total of 17.2 million HIV cases in the world, which had resulted in 2.9 million deaths up to the end of 1994 (Bongaarts 1996: 23). The prevalence of the forced migrations in these areas (discussed earlier in this chapter), in concentrating the movements of infected people in space and time, exacerbates the situation, and the chaotic conditions in destination areas do not allow the ready implementation of the necessary public health measures. Malaria can be transmitted from areas which are malaria-endemic to areas that are malaria-free, and people from malaria-free areas may flee, or be forced to flee, to areas where it is endemic. See Prothero (1994) for a comprehensive review of the recent evidence.

The deteriorating areas, much more than in any other tier, are facing crises, economic, political and in terms of health, and have seen a reversal or at least stasis in migration to the urban sector and other nuclei of commercial development but an increase in forced displacements. The patterns of mobility are reinforcing a de-linking of rural and urban both within and between tiers, in total contrast to the situation in other tiers.

'Smallness' and migration and development

There is no fixed definition of 'smallness'. Many of the issues discussed earlier in this chapter have been aggravated by smallness: the size of ethnic groups in eastern Africa or the size of the elite groups, for example. Yet, in a world of 209 nation states and territories, 61 have populations of less than 1 million, 49 of which have less than half a million people. A further three countries have populations just around the 1 million mark. While size is an unreliable development indicator, with some small nations being very prosperous, all are economically and politically weak within a global system.

Unlike the areas just discussed, the small nations are intimately linked to core or expanding core areas through population migration and thus have much more in common with the labour frontier.

Neverthess, their smallness renders them vulnerable in a way that is not common in the labour frontier. The emigration from a few has been so intense that their demographic viability has been undermined, and the majority are economically and politically dependent upon core areas for their survival as functioning entities. The two main areas of small countries are the Pacific and the Caribbean and, although there are profound differences between them, there are also important similarities. Between 1985 and 1990, the Cook Islands, Niue and Tonga in the Pacific and Aruba, Grenada and Saint Kitts and Nevis in the Caribbean all recorded negative population growth. Between 1990 and 1995, Dominica and Montserrat joined the ranks of those with declining populations, although some of those which had earlier declined showed renewed, if slow, growth.

Migration is the primary cause of the low or negative growth. More than 6 million people left the Caribbean from the end of the Second World War to the late 1980s (Simmons and Guengant 1992: 94) and more islanders live overseas than in the islands themselves in several of the microstates of the Pacific (Connell 1990b: 1). The principal destinations of Pacific migration are New Zealand, principally Auckland, which has become the largest city of Pacific islanders in the world, Australia and the west coast of the United States. Although the Pacific islanders have always engaged in complex systems of circulation, both within and between islands, the 'mass' migrations, in Pacific terms, did not occur until after the changes to core country immigration policies from the 1960s. There are major differences in the migration behaviour between the three major ethnic groups in the Pacific: Melanesians, Micronesians and Polynesians. The former, concentrated in the larger islands in the western Pacific, rarely participate in international migrations but in increasingly complex systems of internal circulation to towns. As in so many other societies undergoing early phases of development, the trend is of a transition towards more long-term circulation or more permanent movements (Skeldon 1990). The Polynesians from the southern and western parts of their vast area move to New Zealand, while those to the north and east, as well as the Micronesians, tend to move to the United States. The most important factor in explaining these migrations is the history of colonial involvement, with the inhabitants of New Zealand's colonies of the Cook Islands, Niue, the Tokelaus and western Samoa moving to New Zealand, and the association of the United States with Hawaii, American Samoa, Guam and the former Trust Territory of the Pacific Islands facilitating movement to the American 'mainland'. Again, much of this movement is circular.

One curious aspect about migration and development in the Pacific is that there are probably more studies of migration per migrant head than for any other part of the world. The reviews by

Bedford (1992), Hayes (1992), Walsh (1992), the massive, multi-volume series by Connell (1983–5) and the edited collections by Chapman (1985), Chapman and Prothero (1985) and Connell (1990a) guide the reader into a vast number of macrolevel and microlevel studies of human mobility. The Caribbean, too, has seen no small number of equivalent studies with the review chapters of Patterson (1978), Pessar (1991) and Simmons and Guengant (1992) providing paths into an extensive literature and the work of Thomas-Hope (1992) and Conway (1994) giving general insight. Small islands in tropical seas have attracted large numbers of outside researchers, and it is perhaps no accident that so many of the conclusions have tended to emphasize the negative impacts of migration on development as self-reliant subsistence economies have been transformed into part societies and part economies. Although a perceived rather than real paradise may have been lost, the analyses of the small island communities can throw light on the process of migration in larger and more complex societies.

In analyses of development in the Pacific, a special term has been coined to describe the nature of the economies: MIRAB or migration, remittances and bureaucracy (Bertram and Watters 1985, 1986). In this interpretation, the islands become dependent upon the remittances sent back by migrants to core areas and upon the aid that the core areas grant to the islanders. Migration is thus an integral part of the very survival of the societies and not a temporary or aberrant phenomenon. The MIRAB interpretation, however, tends to assume a domestic economy in stasis and unchanging demographic variables that guarantee a continuous supply of emigrants to maintain the flow of remittances. It also suggests that local development initiative has been sapped by the emigration. More recent research (Brown 1994, Brown and Connell 1993, Brown and Connell 1995) has shown not only that remittance income has been underestimated but that it can create many opportunities for investment, savings and local development. In particular, the transfer of remittances in kind leads to thriving informal markets and the establishment of a business environment in what Brown and Connell (1993) term the 'global flea market'. Hence, the emigration can lay the basis for forms of development. Nevertheless, the emigration from the islands does appear to drain them of their most highly educated young people and to increase wealth differences in the home villages between those who have successful kin overseas and those who do not. Also, as Hayes (1992) has made clear from studies in Polynesia, declining fertility has reduced the potential future supply of emigrants, undermining the long-term viability of the MIRAB economy.

Although the emigration may stimulate business development in the villages, the loss of young men and women, and their reluctance to engage in traditional activities upon their return, have led to the decline of the agricultural sector. Connell (1994) has shown how the peoples of Micronesia have become heavily dependent on external sources of food supply, by as much as 90 per cent in the Marshall Islands, and acknowledges that no one there would wish to change this situation. While this is perhaps an extreme case, dependency on external supplies of the most basic of essentials is a common pattern throughout the Pacific. These small economies have, to a large extent, lost their *raison d'être* and the Pacific, were it not for its resource endowment of value to core countries, would in another 100 years be an 'empty quarter' (Ward 1989).

The Caribbean has experienced a very different migration and development history. Suffering a demographic collapse of the indigenous population after initial contact with the Europeans from the late fifteenth century, the Caribbean was repopulated by African slaves from the second half of the seventeenth century. The abolition of slavery in the nineteenth century brought about adjustments in population as freed slaves moved between islands or moved into the interior of islands to establish smallholdings. Migration as a 'weapon' was institutionalized in Caribbean societies in a way not seen in the Pacific, whereby labour was withdrawn from the dominant group through flight (Patterson 1978: 127), although such a 'weapon of the weak' has been recognized in many other societies (Scott 1987). A tradition of inter-island mobility has persisted until today and there are much higher proportions of foreign-born among the Caribbean populations than among the island nations of the Pacific. Only in Fiji was there a comparable situation, with large numbers from a different ethnic group being brought in to work on plantations, though small numbers of Chinese are also found throughout the Pacific and the Caribbean.

Although colonial ties were important in explaining the migration out of the Caribbean, proximity to the United States and to opportunities in Central and South America complicated the flows. Migration from the Anglo-Caribbean to the United States and to Britain tended to be countercyclical (Peach 1995). The inhabitants of these areas could enter the United States under the British quota during the first half of this century but when that opportunity was removed, from 1952, the migration turned to the booming postwar British economy, until the British restricted immigration from 1962. The change in the United States immigration act from 1965 rechannelled Caribbean migration back to North America.

Despite the profound cultural and historical differences between the two areas of small nations, there is also much in common. Both have strong cultures of migration in which movement away from the home has become a *rite de passage* for one reason or another. Both have become dependent upon external sources of remittances in cash or kind. Both are dependent upon small amounts of foreign skilled labour as their populations are often too small to generate the necessary division of labour to participate in a global economy. Agricultural production appears to be in crisis in the Caribbean, too, as the women left behind fail to find suitable labour for their smallholdings or the migrant families engage in land speculation instead of planting (Pessar 1991: 207–8). Circulation to core areas is an integral part of the total mobility in both areas. Both are likely to be sustained over the longer term, however, by another form of short-term mobility that brings prosperity to some but increased dependence upon core countries: tourism.

Tourism has become the second largest element in international trade after oil and, although these two areas account for only a minuscule proportion of the total tourist industry, the impact on the small economies and societies is profound. Over 2 million tourist arrivals were registered in Melanesia, Micronesia and Polynesia in 1992 and some 11.7 million in the Caribbean, which together represented less than 3 per cent of total tourist arrivals worldwide (WTO 1994). The small nations can capitalize upon their natural environmental assets, although these will need to be conserved in a way that will sustain the industry into the future. Annual average rates of increase in tourist arrivals between 1980 and 1992 of 10.3 and 4.7 per cent for Micronesia and the Caribbean respectively may not be sustainable, even over the short term, without serious damage to the environment. Receipts from tourism are growing even faster at 19.9 and 8.9 per cent respectively for those two areas and will ensure that there is pressure for the momentum to continue. Core country interests tend to dominate the ownership of the industry and, although there are unquestionably local multiplier effects in terms of employment and the supply of some local produce and handicrafts, this type of population migration will lock these areas into their position as a resource niche.

Smallness is relative. This discussion has focused upon the small island nations of the Pacific and the Caribbean, leaving entirely to one side the question of whether larger countries such as Papua New Guinea or Jamaica should be included, but simply accepting that 'we know a small country when we see it' (Streeten 1993: 197). Clearly, many other areas could be considered under the same category: small, isolated populations in mountainous environments in the Himalayas or the Andes; the shifting tropical rainforest populations of Amazonia or the island of Borneo; the nomadic herding peoples of the Tibetan Plateau and much of Central Asia. Many of these people have turned their backs on dominant groups to continue, as far as possible, their traditional patterns of mobility. These are the 'regions of refuge' (Skeldon 1985), which will be allowed to maintain their ways of life until core areas require resources which are to be had locally. Whether these are minerals, water or simply the delights of an 'unspoilt' environment for tourism, the result will be the same: the young people will be incorporated into the expanding modern sector, populations will atrophy and decline, and the areas will be maintained by temporary migration from, and as outliers of, the core areas.

Developed underdevelopment: the curious case of the Gulf

It may seem strange to argue for the inclusion of the oil-rich states of West Asia in the same development tier as some of the weakest and poorest areas in the world. All the Gulf states rank either among the high-income or the upper-middle-income economies in the World Bank ranking. Yet, in terms of population and migration, they exhibit many of the issues discussed earlier writ large. With the exception of Saudi Arabia, they are all small states with populations of less than 2 million. Unlike the small nations discussed above, all have sizeable migrant populations both absolutely and relatively. The proportions of immigrants range from an astonishing 90.2 per cent for the United Arab Emirates, representing almost 1.5 million migrants, to 25.8 per cent for Saudi Arabia, accounting for just over 4 million migrants (United Nations 1995b). As much as two thirds of the world's known reserves of oil are to be found in the ground around the Gulf and, from 1973, the small states of the Gulf, together with the other members of OPEC, increased the cost of their product by 70 per cent. The resulting dramatic increase in revenue ignited a development boom that could only be supported through the importation of labour to augment the small populations of the Gulf.

The required labour initially came from neighbouring Arab countries, especially Egypt, Jordan and Yemen, which became 'arguably the most "migrant dependent" countries in the world' (Findlay 1994: 103). The demands for labour were such, however, that they could not be met locally and the Gulf states turned to more distant sources: first to South Asia and later to Southeast and East Asia. The origins and impact of these migrations were discussed in Chapter 6. The labour was primarily in the construction industry as the Gulf countries embarked on a frenzy of building a new infrastructure of roads, ports and airports, as well as schools, colleges and administrative blocks, symbols of the new wealth. Many of these schemes were misguided, with much duplication and waste, such as a drive for agricultural self-sufficiency which resulted in the production of wheat and dairy products in the desert at a cost many times more than on the international market and which took a heavy toll on the environment.

The real long-term development benefits of the oil price boom for Gulf states are questionable. These states are, in many ways, caught on the horns of a dilemma. In order for them to achieve a self-sustainable development, an industrial superstructure needs to be built on top of the infrastructural base. In a world of global communications and air cargo, which liberate countries from the constraints of local markets, this should be an achievable objective, given the accumulated wealth of the states. The establishment of assembly industries is theoretically every bit as possible around the shores of the Gulf as along the Mexican border with the United States or in southern China. The critical questions, however, refer to labour and to political stability. In order to establish such industries, a stable, guaranteed labour force is required. The arguments on the changing skill requirements and labour force in South Africa were discussed in Chapter 5, and concern has already been expressed in the Gulf states about the negative consequences of rapid turnover in foreign labour (Shah 1995: 1010). However, it is one thing to import large numbers of labourers for limited-life projects in construction and entirely another to bring workers in for recurrent production in key areas of the economy. The tendency for guest workers to turn into permanent settlers would pose severe difficulties in conservative Islamic societies such as Saudi Arabia, particularly if the young women required in modern light industrialization were to be recruited and were to bring family members from non-Islamic areas. Large numbers of temporary workers already raise questions within a factionalized Arab society, which would certainly be exacerbated by the presence of a permanent foreign industrial labour force.

The answer would obviously be to depend upon local sources of skilled labour, which would be employed in highly automated factories to compensate for the small size of the local labour force, and these skilled workers could be supported by short-term skilled advisers from core countries. However, the indications are that the skill levels of local labour in Gulf states have not improved significantly and that, rather than take up 'a taxing private sector job, locals engage in rewarding (and less demanding) activities in which they can charge commissions as sponsors (*kafeel*), or take desk jobs in the government bureaucracy' (Shah 1994: 14). Thus, the alternatives appear to be either to achieve sustainable development through a permanent foreign labour force, with the risk of the erosion of cultural or even national integrity, or to adopt a less developmental but culturally more acceptable strategy. Perhaps the simplest form of the latter is to invest overseas for the benefit of the citizens of the Gulf states, and particularly the rulers. Kuwait has been a long-term investor on international markets to the extent that, when oil prices began to fall from the mid-1980s, returns from investment began to overtake oil revenue and it became, in Findlay's term, the first 'rentier' economy in the world (1994: 98). The United Arab Emirates, and to a lesser extent Saudi Arabia, have also followed this strategy, with the economic base of these states now shifting to core areas.

Were the oil reserves to become exhausted or the core areas' thirst for oil to be moderated, neither of which appears remotely probable in the near term, then the small populations of the Gulf states would be able to survive on their investment income, although presumably not the migrants, who would return to their home countries. More likely is the possibility of regional unrest as was demonstrated in the 1990–1 Gulf conflict. Within two months of the conflict, some 900,000 workers left Kuwait and 275,000 left Iraq 'in one of the largest and most rapid repatriations in recent history' (United Nations 1996: 209). While Kuwait has replaced its foreign workers to at least pre-conflict levels, their composition has changed. The exodus of Palestinians and Jordanians from Kuwait is likely to have been permanent, owing to their leaders' sympathies during the conflict, while Egyptians and Asians are likely to have been favoured in the post-conflict recruitment. The point is clear: the migrants are in an extremely vulnerable position, and future economic or political shocks can return the Gulf states to small populations in harsh environments without the local basis for sustainable development. These areas, too, would then become regions of outmigration to core countries. While this scenario is largely hypothetical, it reinforces the vulnerability of populations and environments in this development tier: the MIRAB economy of the Pacific has its dependent equivalent in West Asia – OMOIB – oil, migration, overseas investment and bureaucracy.

Discussion

The component parts of this development tier indeed show great variation in their patterns of migration and development. There are no trends towards any convergence or common patterns of mobility, save that circular, non-permanent forms tend to dominate. Common to the tier are weak state structures that are the result of the colonial imposition of boundaries, amalgamating often disparate ethnic groups. Centralized systems of mobility that might provide substance to a national identity tend to be lacking or to be at levels below that of the existing state. Secessionist pressures tend to be strong and the majority of states are weak, if not small and weak. Conflict between competing local interests generates refugees. The retreat of the labour frontier, the reversal in rural-to-urban migration in parts of Africa and the exodus of Russians from Central Asia further undermine the linkages with other development tiers, exacerbating the isolation and weakness of this tier. It appears that few external forces at present are willing to foster unity except where resources exist of critical value to core countries. The exceptional commodity, oil, and the past ability of local groups to increase their leverage over its pricing, caused a unique and massive temporary labour migration into one part of the tier. The establishment of a more broadly based development, however, appears unlikely, as this would imply an institutionalization of immigrant groups through long-term settlement that would be unacceptable to local leaders and local populations. No parts of this tier are now open for the establishment of settler societies, in part because of environmental constraints but mainly due to political will.

Large parts of this development tier have weak linkages with the other tiers or have been marginalized, or have marginalized themselves, from the global system. Where the linkages are stronger, large-scale depopulation of all but service centres seems a long-term possibility, with the populations shifting to core settler countries. Where they are weaker, the areas will be left virtually to their own devices, until some resource is discovered of value to the core countries. There may be proactive resistance towards incursions from outside and attempts to exclude influences from other tiers, but these, over the long term, are unlikely to prevail, given the small populations and weak economic organization.

Where resources required by the core countries are discovered, the size of the resultant immigration will depend upon their nature and value but, in this tier, it appears unlikely that any truly sustainable and long-term development will result. The negative impacts of migration through the brain drain and the diffusion of disease are much more a feature of this tier than elsewhere. All that will exist over the long term will be a series of urban nuclei set in a virtually depopulated environment and staffed by a few remaining local inhabitants supported by skilled inmigrants from the core countries to oversee operations, whether in extraction or in recreation. This scenario does not yet typify the whole tier, of course. There is still enough labour in the labour frontier and little need to incorporate the populations in the more densely settled areas in the resource niche. Nevertheless, future physical resources will almost certainly be exploited in this tier as sources in more accessible areas, both politically and geographically, are exhausted. This tier is both the present and the future resource niche of core areas and, within a global system, it is difficult to foresee any dynamic growth in the tier or any major centres of long-term inmigration.

Conclusion: the system and the future

The system

The first substantive issue that should be clear from the discussions of migration and development in the five development tiers is that population movements are an integral part of the societies and economies of each one: the forms and type may differ, but people are highly mobile in each tier. The examination of the evolution of mobility should also have shown that the forms change over time and that it would be difficult to identify a time when migration was unimportant. Our era is certainly *an* age of migration, but it is not *the* age of migration. There have been many times in the past when migration has been just as significant. Although more people may be moving over longer distances today than ever before, this is due primarily to the fact that there are more people alive in the world today and that modern technological development in transport has facilitated the conquest of distance.

The second substantive issue that I hope has emerged from these discussions is that migration is still largely regional rather than global. Although we are in an era of global and virtually instantaneous capital flows, the networks of both migration and capital are highly concentrated. The first three tiers are closely bound together in terms of capital and population flows, but the last tier, the resource niche, is only tenuously linked to the other tiers. The number of migrants moving out of that tier is still quite limited or, where it is relatively important, the populations are small on a global scale. There is no massive uncontrolled movement out of the poorest parts of the world towards the richest parts of the world. What has occurred is a tremendous increase in movement, but that movement tends to be constrained mainly within each country at lower levels of development, or certainly within each tier. Where there is increasing migration into the most prosperous core countries, it tends to come from adjacent areas and these are rarely the poorest areas. Regional systems of migration have evolved and these are intensifying: much of this is distress movement of one type or another and much is movement to local and regional urban centres.

Some migration fields are global in their extent: multinational corporations and international agencies that are based in core countries; the United States that takes in settlers from every part of the world; and a few ethnic networks, such as those of the Chinese, that are of truly global dimensions. Nevertheless, the intensity of the movements within each of these systems tends to be greatest among the core and expanding core tiers themselves, or from the interface of the labour frontier and the cores.

Whether migration is seen as positive or negative for development will depend very much on the context. Migration both results from change and engenders further change. Once we accept that migration is an integral part of the behaviour of all societies at all times, and is not 'abnormal' or exceptional, then the objective is to see how its form and function are related to other change in economies and societies. There are transitions in mobility, but not in the sense that all societies are moving uniformly through a sequence of stages from one state of stability, through a period of flux, to another state of stability. Rather, the transitions are constant shifts in the pattern of mobility that are related to other broad changes in economy and society. These shifts are systematic but not uniform in time and place. We have seen that, in the core tiers, despite differences in mobility rates, there has been a convergence in the patterns of movement over time. At the lower end of the development spectrum, the resource niche, there were few common trends and a great diversity of migration types and patterns. As we moved from the old to the new core to the expanding core tiers, the time period for the transition from rural to urban, from a labour-surplus to labour-deficit economy and from a high-fertility to a low-fertility society was reduced. There were several paths to an urban-industrial economy, with distinct parallels, but also important variations. International movements out of the new core were never as important as for the old core areas in Europe, mainly because migrants from the new core were moving into environments controlled by or at least heavily influenced by migrants from the old core. Later migrations are always conditioned by earlier migrations and policy can be an important constraint, as we saw in the case of the exclusion of Asian migrants from North America and Australasia.

The third substantive issue that should be clear from the discussions is that the most developed societies, how ever defined, are predominantly urban. The city is the defining element of a civilization in more ways than the philological, and a transition to an urban society and the concentration of population are key elements of human progress. Cities provide the marketplace for exchange and the creation of wealth. The bringing together of peoples from different backgrounds, often in polyethnic communities, has accumulated the talented and the most able members of populations. Very often the migrant, or a member of a minority community, has been a catalyst for change, introducing new ideas and ways of doing things. Of course, it is not the migration that creates the talent; either the talented may be made scapegoats and forced to migrate, or the isolated position of the migrant thrusts him or her into a situation where they must innovate to survive: the refugee mentality. From the Huguenots to the Jews to the East African Indians and the Hong Kong Chinese, migrants have been in the forefront of entrepreneurial and intellectual development.

The process of concentration shows similarities as well as differences from one part of the world to another. Clear parallels can be seen between the urbanization experiences of Europe and East Asia (Rozman 1990), that is, the two core tiers. The progression from systems of circulation through longer periods of residence in the largest cities and permanent transfers of population during a period of intense concentration, to the deconcentration of populations once high levels of urbanization were attained, were observed in these tiers. The rapidly expanding core has seen the shift from systems of short-term circulation to the present intense concentration and semi-permanent migration to the largest cities.

Migration has a variable impact on the communities of origin. Initially, it acts as a support for rural communities, extending the resource base of the domestic economy. Remittances improve the level of living for the majority of households with migrant members. Yet, as migration leads to further migration and the demand at destinations is for a more stable labour force, the rural communities are gradually drained of their populations. Once they lose their populations of reproductive age on a semi-permanent basis, the communities enter a period of permanent decline. The succeeding generation is born in town and, even if the parents eventually return to the village to retire, the children have their roots elsewhere and have little interest in the rural sector. In areas of favourable location and physical environment, the removal of population can be positive, leading to the commercialization of agriculture and rise of real income in the way foreseen in the classical models of migration and development. In less favoured areas, rural depopulation is the outcome.

One of the most significant shifts in thinking about the process of migration and development in recent years has been 'an acceptance that urbanisation in developing countries is *inevitable*' (McGee 1994: iii, emphasis in the original). This statement comes from a leading authority on urbanization who long argued that the pattern of urbanization in less developed countries was going to be very different from that of the developed world (for example, McGee 1991). One of the criticisms levelled at generalizing models such as the mobility transition (Zelinsky 1971) and my variation of it (Skeldon 1990) was that they argued for the inevitability of population concentration in large cities, possibly followed by some deconcentration at advanced stages. While it might appear to be gratifying to see a more general acceptance of these conclusions, what the review in the chapters in this book has clearly shown is that there is nothing inevitable about urbanization in all parts of the world. In those areas where rapid development is occurring, it indeed appears inevitable, but there are also large parts of the world where urbanization is slowing, even reversing, and others where no independent or integrated urban system will develop and depopulation is a more likely outcome.

Whether these deviations are merely temporary blips in a long trend towards a global urban society or an indication of a separate pattern is obviously of significance to this debate. I argue that the value of a regional approach, such as that adopted in this book, is that it helps us to see where and how the urban growth has taken and is likely to take place. In effect, it can be envisaged as a modification of dependency theory, in which the simple bipolar core–periphery division is replaced with a gradation of tiers in which the boundaries can shift and some parts can attain a self-sustaining development, while others are locked into dependency and are unlikely to achieve acceptable levels of development.

Nevertheless, there are problems with the regional approach. As emphasized in Chapter 2, the issue of boundaries is fuzzy. There could be much discussion over whether central Chile, or the Jakarta urban area, for example, should be included as potential cores instead of in the labour frontier. Clearly, McGee's (1994) scenario for Jakarta in the year 2020 identifies that region as part of a future core. As the capital region of one of the largest countries in the world, and one rich in natural resources, it is certainly one of the areas that should have potential. However, given the existing labour surplus of Indonesia, the fact that migration into the Jakarta metropolitan area is still overwhelmingly from elsewhere in Java, even if there is indeed growing ethnic diversity, and the absence of institutions for the smooth transfer of power, this area is, for the moment, not included in the potential core tier. Indonesia is still more a sending than a receiving country as far as migration is concerned, and one of the critical features of all core and expanding core areas is their ability to act as significant poles of attraction in regional and global migration systems.

One could also cavil about the inclusion of Ireland in the old core tier: it may be more appropriate as part of the labour frontier. Perhaps the very intensity of the emigration over the last 150 years, by removing the 'surplus' population that might have agitated for economic and political change, actually condemned Ireland to a permanent peripheral position as a labour frontier within Europe

(MacLaughlin 1994). There have, however, been fluctuations in the overall trend of emigration. During the decade of the 1970s, Ireland experienced a 'sustained net inflow of population unknown since the republic gained independence in the 1920s' (Almeida 1992: 197). Its development, at that time, was stimulated through the establishment of multinational companies. These, however, incorporate their employees into international career paths. When there is an economic downturn, these employees are likely to be transferred to more profitable locations. Thus, the factors that have brought development and retained potential emigrants, even brought others from overseas, can also facilitate outmigration when economic conditions change. Ireland saw an exodus of middle-class migrants in the 1980s, many of whom were university graduates educated by the state (Hanlon 1992), and the net flows again became negative at −208,000 for the decade. Not all the new emigrants are highly educated but many, perhaps the majority, come from low-income groups in small towns and rural parts of the country. Ireland, it can be argued, still retains its role as an 'emigrant nursery', supplying core areas with labour for unskilled, low-paid and often temporary jobs (MacLaughlin 1994). The majority of the 'new Irish' migrants left for Britain, although many, too, became illegal migrants in the United States (Almeida 1992). However, Ireland, with its GNP per capita in 1993 at US$13,000 and ranked nineteenth out of 174 countries in the UNDP human development index in 1995, demonstrates that a labour frontier close to the core can attain high levels of development.

The above example draws attention to an issue that has so far received scant attention in this global assessment of migration and development. We have seen that there were labour frontiers close to and within the core tiers themselves: Ireland, Portugal, southern Italy and eastern Europe have all been mentioned. However, there are also resource niches within the core tiers. These are areas that, in the history of capitalist expansion, were drained of their population and are now recreation areas or supply some other resource (hydropower, for example) for the urban industrial economy. Figure 3.2 (page 74), showing the areas of net outmigration in Europe, gives an idea of their extent in that part of the old core tier. The areas of intense depopulation in Japan, *kaso* (Skeldon 1990: 119), provide the counterpart for the new core tier. In both core tiers, the mountainous and marginal parts have been abandoned to all but service workers and affluent retirees from the core cities in a historical counterpart to the present and future courses of migration in several parts of the resource niche described in Chapter 7.

Thus, an archipelagic pattern of development tiers might be more exact than the broad bands depicted in Fig. 2.2 (page 51) although, at the global level, the resulting pattern might be too complex to comprehend visually. In addition, to include in the tier identified as the resource niche the recreational and resource areas within the core is to ignore one of the critical development criteria, the efficacy of the state. Core tiers, by definition, are characterized by strong states and, as we move down through the tiers, these structures become progressively weaker. One of the key factors in explaining mobility in the resource niche was the inability of the state to guarantee the protection of the populations within its territory, which is quite different from the situation in the core tiers. Admittedly, some groups within the core tiers can be seen as encapsulated cultures in developed societies. First Australians or Canadians fall into this category, being groups whose indigenous development has been eroded as the immigrant groups have progressed. Their mobility is quite different from that of the urban-based settler majorities, although that distinctiveness is also due to the provisions of liberal welfare states. In Australia, the aboriginal population has apparently returned to increasingly remote areas, facilitated by more widespread acceptance of the importance of indigenous rights, to pursue an independent lifestyle supported by state welfare (Taylor and Bell 1996). Minorities in the resource niche are often persecuted, as we saw in Chapter 7, or subject to intense pressure to adopt the values of the dominant society in what is, in effect, internal neo-colonialism. The 'Thai–ization' of the hill tribes of northern Thailand or the sinification of Tibet and western China would be examples of such approaches.

Such approaches were, of course, hardly absent from the history of the core tier countries. The incorporation of the populations within a delimited territory is a prerogative of state formation everywhere, and mobility is fundamental to this development. Perceived uncontrolled migration among peripheral groups suggests that they are not under state control. They may even move backwards and forwards to neighbouring states, indicating that they need to be integrated into state institutions. The role of the military has been raised at several points in the discussion, and the centralized circuits of recruitment are important in incorporating peripheral groups into central authority. Education, also initially concentrated in the larger towns, is another powerful centralizing and state-building force. The inevitability of urbanization observed earlier is intimately related to this process of state formation and explains the lack of success of policies designed to keep migrants 'down on the farm', and of those encouraging decentralization. The state is essentially defined through centralized circuits of population mobility and will not actively promote policies that might conflict with this objective. Only once the state has reached a stage of considerable maturity and self-confidence in the systems for the transfer of power can it afford the luxury of promoting policies of decentralization. Before this stage is attained, decentralization is likely to lead to secession and the possible dissolution of the state, as seen in some of the parts of the resource niche discussed in Chapter 7. Thus, the regional approach to development, though combining different economic zones within a single tier, seems a reasonable compromise as a heuristic device as it combines broad structural differences in economic development with an indication of the effectiveness of the state.

A critical substantive issue is to what extent the cores can continue to expand. In the past, the boundaries between the tiers shifted markedly, particularly as the labour frontier pushed ever further into the periphery and as parts of the original labour frontier were transformed into cores. Will the capitalist system extend, as we saw in Chapter 5, ultimately to incorporate the whole world and create a truly global system? Not that the cores themselves will remain static, with a uniformly developed and prosperous world the end result. It has been an important part of the argument in this book that the core tiers are also developing, and developing faster than the majority of countries in the labour frontier or resource niche. Differences in relative wealth are not going to disappear and will almost certainly continue to increase, even if some areas may emerge as new poles of growth in a way similar to that described in Chapter 5 for the rapidly expanding cores. The boundaries of the tiers are certainly flexible and it is perfectly possible for regions such as the Jakarta metropolitan area to become a major regional growth centre, attracting migrants from south and east Asia and further afield as local sources of labour diminish. Yet, the evidence presented in Chapters 6 and 7 suggests that these new future growth poles may be quite limited in number. Urbanization may be inevitable but the urban centres will be concentrated in relatively few areas, and the rest of the world will gradually be drawn into their ambit.

There appear to be some strong centralizing circuits of mobility in the labour frontier, even if much is directed to centres outside the tier, so the possibility for the emergence of strong states certainly does exist in this tier. Chapter 6 ended with a discussion of such possible states. Interestingly, the majority of the states recognized as 'pivotal' by Chase, Hill and Kennedy (1996) include part of an expanding core tier, as identified in this book, or lie on the interface between the labour frontier and core tiers. In the resource niche, there appear to be few strong centralizing population movements: a profusion of complex and changing flows exists within the tier. Where there is migration out of the tier, societies and economies may eventually be undermined by the loss of the elite or, in small economies, the reproductive capacity, leading to population decline. Strong states are unlikely to emerge in this tier and, when the resources of the tier are required by the cores, the populations will be drawn into the migration fields of core cities, leaving these areas in permanent dependency. Thus, although there may be a convergence in the spatial patterns of migration in the core tiers and a profusion of different types at the opposite end of the development spectrum, the differences between the latter would diminish as they were incorporated into urbanward flows to core areas. That incorporation, however, may be slow to come about and, in some areas, may never happen.

The future

As we move towards the twenty-first century, migration will indeed be a major issue of public and academic concern. The decline of fertility is now either complete or well under way in virtually all major regions of the world. Even in sub-Saharan Africa, there is now evidence of the beginnings of a sustained decline. Together with ageing, essentially the result of the fertility decline, one of the principal concerns in the general subject area of population studies will be migration and population redistribution. There are clear interrelations between the three demographic variables. The important cohort effects on migration were referred to in Chapter 3 and ageing and the slowing in the rate of growth of the labour force have profound implications for international migration, as we have seen in several chapters. Clearly, demographic factors are not the only, or necessarily the most important, variables in accounting for migration, yet they must be part of the equation. It is curious that so little effort has been made to examine the relationships between them to elaborate a truly integrated model of the demographic transition. Although Zelinsky (1971) proposed such a relationship, just how the variables interacted was never specified.

International agencies have been ambivalent in their approaches to migration. Only the International Labour Organization (ILO) has consistently held a high profile in the area, although focusing mainly, and understandably, on labour migrants and their rights. The chief United Nations organization for population, the United Nations Population Fund (UNFPA), has rarely made migration a priority. In part, this is due to the multidimensional nature of migration, which places it within the purview of several agencies, with the resultant possibility of duplication. In mid-1996, the United Nations Educational, Scientific and Cultural Organization (UNESCO), the Population Division of the United Nations, the Organisation for Economic Co-operation and Development

(OECD) and the International Organization for Migration (IOM), as well as the ILO, all had programmes to examine or support activities in migration, although almost all of these programmes were directed towards international migration. In part, however, the ambivalent attitude of international agencies is also due to the peculiar policy implications of migration. Internal migration has proved singularly intractable towards policy intervention. Once population movements emerge, they take on a momentum of their own that is virtually impossible to control, except by governments of the extreme right or extreme left. The pass laws of South Africa under apartheid, the household registration system of Maoist China and the de-urbanization policy of Pol Pot's Cambodia are three of the few policies that have controlled migration effectively, although few would recommend that they be used as models. As emphasized in the Introduction, freedom of movement is a basic human right in the strong states of the cores, and any attempt to modify that right would be seen to be contrary to human development. As far as international migration is concerned, the control of national borders is seen as the responsibility of each national government, which again does not lend itself to easy policy intervention from transnational agencies. All the indications are, however, that there is a convergence in policy approaches in core tier countries, and that convergence is towards restriction of entry (Chapter 3).

The reaction leading to these restrictive attitudes and policies is largely alarmist. We have already seen that most of the increasing population movement is largely contained within tiers and within countries. For example, the vast majority of the estimated 9 million former Soviet citizens who have moved since 1989 have not crossed over to western Europe but have migrated within the boundaries of the former Soviet state, and we have seen the emergence of regional systems of migration in Asia, Latin America and Africa. The idea of a South–North migration of unprecedented magnitude is not only geographically incorrect, but it also obscures the most significant flows. The migration that does occur towards the main settler countries, although large absolutely, is not particularly great relatively by historical standards. Much of the reaction against migration is engendered by fear: a fear that the familiar homogeneous nation state of the last 250 years of western history is giving way to something different and more complex. There is the 'reassertion of the polyethnic norm' (McNeill 1986). Clearly, this reassertion applies principally at the upper levels of the development hierarchy whereas, at the lower end, the idea of the homogeneous state is creating much migration. At the bottom end, economic and political forces favour expulsion whereas, at the top, they favour attraction.

The major issues in the area of migration, as far as international agencies are concerned, will probably be illegal migration and human trafficking. These activities require international cooperation and have taken on an importance that rivals the trade in drugs, to which it is closely related. Organized criminal syndicates were smuggling perhaps 100,000 people from China into the United States alone in the early 1990s, and thousands more were smuggled from eastern Europe and Russia into western Europe. The trade in human beings, as old as history, persists into the most advanced societies, representing a tragic and contradictory link between migration and development.

The numbers entering a destination should not be the main consideration but the composition of the migrant flows. A small number of entrepreneurial, highly educated and ambitious migrants are likely to have as great an impact on the destination society as large numbers of poorly educated migrants, and a very different kind of impact. The main settler societies have been able to attract, to a variable degree certainly, some of the best-educated, as well as the poorer workers. As we saw in Chapters 3 and 4, the inmigration and any return migration had profound effects on the course of development of origin and destination societies. The net benefits to both societies appear to have offset any short-term costs that may have been incurred. Immigrants tend to create more wealth than they consume and return migrants are important contributors to more open societies. Even if attempts are made to limit immigration, these are likely to be shortlived in the core tiers, given the pressures of tight labour markets. In order to remain competitive in a global economy, nations need access to the best brains, and the mix of peoples and talents brought about through migration will always give the key poles of attraction, the global cities, an advantage.

This book has focused upon the macrolevel interrelationships between migration and development at the global level, with migration considered an essentially social phenomenon. Yet, these population movements are made up of millions of individuals, each with their own experiences of migration, their hopes and aspirations. The social scientist is often uncomfortable with the wealth of individual detail, but approaches that might broadly be classified as postmodernist are providing valuable insight into the nature of population movement (see Chapter 1) – migration as an 'ontological experience' (Ahmad 1992: 13). The topic is approached primarily through the analysis of literature, and the themes of the expatriate experience, of exile, of belonging and non-belonging, and of identity are writ large. See the essays in King, Connell and White (1995), for example, and particularly the introductory chapter of White (1995). Clearly, these migrant experiences are restricted to the literary elite, who, by definition, are very different from the majority of movers. Ahmad (1992: 207–8), for example, takes Said to task over his claim that all migrants are thrust into a position of 'adversarial internationalization', while the reality is that the vast majority, students excepted, identify with the political values of the core country and play no part in an anti–imperialist struggle. In order to comprehend the experiences of the majority, oral histories or biographical approaches have been suggested (see, for example, Halfacree and Boyle 1993). Sensitive anthropological accounts such as that of Wood (1984), which describes the experiences of a group of, mostly illiterate, Indian villagers as they venture out into the outside world for the first time, are unfortunately all too rare.

The focus on the experience of movement certainly broadens the consideration of migration to include all forms of population mobility, including travel. The idea of travelling cultures (Clifford 1992), or how people leave and arrive at places, helps to shift our gaze away from the fixed nature of community towards a more fluid conception of life. Travel literature was critical in the evolution of the modern novel (Adams 1983), that western medium that blends realism and romanticism and tries to capture the essence of the human condition, and thus strengthens the link between literature and the analysis of migration. In the words of Salman Rushdie in *Shame* (see Ahmad 1992: 154), 'we are all migrants' and, although he is clearly making a metaphorical statement, this draws attention to the universality of the experience of mobility. We all move during our lifetimes and life is very often cast as a journey. To be sure, some individuals and, more importantly I would argue, certain groups, move further and more often than others. Nevertheless, it is surely the universality of the migratory experience or, more correctly, the mobility experience, that lends such fascination to the study of population migration. This book has attempted to provide a framework in which to view, and I hope comprehend, the variety of this experience.

Let us leave these ethereal heights behind and return to the concrete. The discussions in this book have highlighted the fact that the relationship between migration and development is not only extremely complex but that it varies by level of development. The relationships in the core tiers may be different from those in the resource niche, for example, and unless the relationships are placed within a specific space–time framework, misinterpretation is likely to be the result. The impact at the macrolevel is distinct from that at the level of the individual. Despite all the differences, the movements of population generate progress and hope: the converse is stagnation. Notwithstanding the fear that so much migration seems to produce in people, the fear of the outsider, it is often the outsider who brings new ideas, some disruptive certainly, which create the tension that leads to economic, social and political development. It is thus almost impossible to envisage development without migration and, although it is a suspiciously trite note upon which to finish, migration *is* development.

Annexe tables

Annexe table 1.   Basic demographic and development variables for countries arranged by development tier

	Population mid-1994 (millions)	Annual growth of population 1990–4 (%)	Annual growth of labour force (%)	Proportion urban 1994 (%)	Migrant stock 1990 (%)	Number of migrants 1990 (thousands)	Remittances as % of foreign exchange earnings	Total fertility rate 1994	GNP per capita 1994 ($US)	Annual average growth of GNP per capita 1985–94 (%)
The old core										
Switzerland	7.0	1.0	1.0	61	16.0	1,092	0.8	1.5	37,930	0.5
Denmark	5.2	0.3	-0.1	85	4.0	211	–	1.8	27,970	1.3
Norway	4.3	0.6	0.7	73	4.4	186	0.2	1.9	26,390	1.4
Germany	81.5	0.6	0.2	86	6.4	5,037	0.4	1.2	25,580	–
Austria	8.0	1.0	0.5	55	5.8	450	1.5	1.5	24,630	2.0
Sweden	8.8	0.6	0.3	83	8.9	761	0.3	1.9	23,530	-0.1
France	57.9	0.5	0.8	73	10.4	5,897	1.2	1.6	23,420	1.6
Belgium	10.1	0.4	0.5	97	9.0	898	–	1.6	22,870	2.3
Netherlands	15.4	0.7	0.7	89	7.8	1,167	0.5	1.6	22,010	1.9
Italy	57.1	0.2	0.4	67	1.1	3,929	0.5	1.3	19,300	1.8
Finland	5.1	0.5	0.2	63	1.2	62	0.1	1.9	18,850	-0.3
United Kingdom	58.4	0.4	0.3	89	6.5	3,718	–	1.8	18,340	1.3
Ireland	3.6	0.5	1.5	57	9.3	326	–	1.9	13,530	5.0
Spain	39.1	0.2	1.0	76	1.8	719	2.3	1.2	13,440	2.8
Portugal	9.9	–	0.4	35	1.4	141	15.1	1.4	9,320	4.0
Greece	10.4	0.6	0.7	65	3.2	322	15.8	1.4	7,700	1.3
United States of America	260.6	1.0	1.1	76	7.9	19,603	0.1	2.0	25,880	1.3
Canada	29.2	1.3	1.1	77	15.5	4,266	0.7	1.9	19,510	0.3
Australia	17.8	1.1	1.6	85	23.4	3,916	2.2	1.9	18,000	1.2
New Zealand	3.5	0.9	1.5	86	15.5	519	7.0	2.1	13,350	0.7
The new core										
Japan	125.0	0.3	0.6	78	0.7	868	0.1	1.5	34,630	3.2
Singapore	2.9	2.0	1.0	100	15.5	418	–	1.8	22,500	6.1
Hong Kong	6.1	1.5	0.8	95	–	–	–	1.2	21,650	5.3

			—	—	—	—	—	—		
Taiwan	20.9	1.4[a]	1.9	80	2.1	900	0.5	2.0	10,852[b]	6.4[a]
South Korea	44.5	0.9	1.9	80	2.1	900	0.5	1.8	8,260	7.8
Core extensions and potential cores										
Malaysia	19.7	2.4	2.7	53	4.2	745	0.2	3.4	3,480	5.6
Thailand	58.0	1.0	1.5	20	0.6	314	2.5	2.0	2,410	8.6
China	1,190.9	1.2	3.7	29	—	346	1.0	1.9	320	2.9
India	913.6	1.8	2.1	27	1.0	8,660	9.3	3.3	530	7.8
Israel	5.4	3.7	3.6	90	30.9	1,427	5.6	2.4	14,530	2.3
South Africa	40.5	2.2	2.5	50	3.1	1,118	2.4	3.9	3,040	-1.3
Mexico	88.5	2.0	2.9	75	0.4	339	7.6	3.2	4,180	0.9
Venezuela	21.2	2.3	3.1	92	5.3	1,028	—	3.2	2,760	0.7
Trinidad and Tobago	1.3	1.2	2.1	66	5.0	61	1.2	2.5	3,740	-2.3
Brazil	159.1	1.7	1.9	77	1.0	65	0.4	2.8	2,970	-0.4
Argentina	34.2	1.2	2.0	88	5.1	1,661	—	2.6	8,110	2.0
Uruguay	3.2	0.6	1.0	90	3.0	93	—	2.2	4,660	2.9
Russian Federation	148.3	—	—	73	—	—	—	1.4	2,650	-4.1
Slovenia	2.0	-0.1	0.3	63	—	—	—	6.3	7,040	—
Hungary	10.3	-0.3	-0.1	64	0.3	30	—	1.6	3,840	-1.2
Czech Republic	10.3	-0.1	0.5	65	—	—	—	1.4	3,200	-2.1
Estonia	1.5	-1.2	-0.4	73	—	—	—	1.5	2,820	-6.1
Latvia	2.5	-1.2	-0.8	73	—	—	—	1.4	2,320	-6.0
Lithuania	3.7	—	-0.1	71	—	—	—	1.5	1,350	-8.0
Poland	38.5	0.3	0.5	64	3.6	1,350	—	1.8	2,410	0.8
Slovak Republic	5.3	0.3	0.9	58	—	—	—	1.7	2,250	-3.0
Belarus	10.4	0.2	-0.1	70	—	—	—	1.6	2,160	-1.9
Ukraine	51.9	—	-0.1	70	—	—	—	1.5	1,910	-8.0
Moldova	4.3	-0.1	0.5	51	—	—	—	2.1	870	—
Romania	22.7	-0.5	0.1	55	0.6	140	0.1	1.4	1,270	-4.5
Bulgaria	8.4	-0.8	-0.5	70	0.2	22	—	1.5	1,250	-2.7

Annexe table 1. *Continued*

	Population mid-1994 (millions)	Annual growth of population 1990–4 (%)	Annual growth of labour force (%)	Proportion urban 1994 (%)	Migrant stock 1990 (%)	Number of migrants 1990 (thousands)	Remittances as % of foreign exchange earnings	Total fertility rate 1994	GNP per capita 1994 ($US)	Annual average growth of GNP per capita 1985–94 (%)
The labour frontier										
Albania	3.2	-0.6	1.4	37	0.4	–	56.8	2.7	380	–
Philippines	67.0	2.2	2.6	53	0.1	38	13.4	3.8	950	1.7
Indonesia	190.4	1.6	2.5	34	0.1	96	0.8	2.7	880	6.0
Viet Nam	72.0	2.1	2.1	21	–	21	–	3.1	200	–
Laos	4.7	3.1	2.6	21	0.4	14	2.9	6.6	320	–
Cambodia	10.0	2.8c	2.3c	12c	0.3	22	–	5.3d	–	–
Burma	45.6	2.2	2.1	26	0.2	100	0.5	4.0	–	–
Bangladesh	117.9	1.7	2.7	18	0.7	800	25.5	3.6	220	2.1
Pakistan	126.3	2.9	3.3	34	6.1	7,272	14.7	5.4	430	1.3
Sri Lanka	17.9	1.3	2.0	22	0.1	21	15.2	2.4	640	2.9
Egypt	56.8	2.0	2.9	45	0.3	176	31.1	3.5	720	1.3
Syria	13.8	3.4c	4.3c	50e	6.6	800	11.3	5.9d	–	–
Jordan	4.0	6.0	5.2	71	26.4	1,112	26.3	4.8	1,440	-5.6
Turkey	60.8	2.0	2.3	67	2.0	1,102	9.6	3.2	2,550	1.4
Tunisia	8.8	1.9	3.0	57	0.5	38	9.0	3.0	1,790	2.1
Algeria	27.4	2.3	4.2	55	1.5	370	1.8	3.7	1,650	-2.5
Morocco	26.4	2.0	2.6	48	0.2	42	23.4	3.5	1,140	1.2
Namibia	1.5	2.8	2.6	36	0.6	8	–	5.1	1,970	3.3
Botswana	1.4	3.1	3.2	30	1.8	22	2.5	4.5	2,800	6.6
Zimbabwe	10.8	2.5	2.4	31	8.0	775	0.1	4.0	500	-0.5
Zambia	9.2	3.0	3.0	43	4.1	325	0.1	6.0	350	-1.4
Mozambique	15.5	2.2	2.7	33	0.1	7	16.0	6.6	90	3.8
Chile	14.0	1.5	2.2	86	0.8	106	–	2.5	3,520	6.5
Bolivia	7.2	2.4	2.6	58	1.0	65	0.4	4.7	770	1.7
Peru	23.2	1.9	3.0	72	0.3	67	–	3.1	2,110	-2.0
Ecuador	11.2	2.2	3.1	58	0.8	78	–	3.3	1,280	0.9

Colombia	36.3	1.9	2.6	72	0.3	101	6.3	2.6	1,670	2.4
Guyana	0.8	0.9	—	35^c	0.4	3	—	2.6^d	530	0.4
Suriname	0.4	1.1	—	48^e	2.1	8	0.1	2.7^d	860	1.8
Panama	2.6	1.9	2.6	54	2.6	61	—	2.7	2,580	−1.2
Costa Rica	3.3	2.1	2.8	49	5.9	177	—	2.9	2,400	2.8
Nicaragua	4.2	3.1	4.5	62	2.1	75	5.5	4.9	340	−6.1
Honduras	5.8	3.0	3.7	47	0.7	34	10.4	4.7	600	0.5
El Salvador	5.6	2.2	—	44^e	1.0	50	36.8	3.8	1,360	2.2
Guatemala	10.3	2.9	3.5	41	0.5	44	10.5	5.2	1,200	0.9
Belize	0.2	2.6	—	52^c	12.5	23	5.5	4.2^d	2,530	5.0
Cuba	11.0	0.8	—	75^e	0.6	68	—	1.8^d	—	—
Jamaica	2.5	0.9	1.6	55	0.8	19	9.1	2.5	1,540	3.9
Haiti	7.0	1.9	1.9	31	0.3	18	—	3.6	230	−5.0
Dominican Republic	7.6	1.7	2.7	64	2.5	174	13.1	2.9	1,330	2.2
The resource niche										
Mongolia	2.4	1.9	2.8	60	0.5	10	—	3.4	300	−3.2
Nepal	20.9	2.5	2.5	13	2.1	401	—	5.3	200	2.3
Kazakhstan	16.8	0.1	0.9	59	—	—	—	2.3	1,160	−6.5
Afghanistan	22.8	5.9^c	6.4^c	18^e	0.2	30	—	—	—	—
Kyrgyzstan	4.5	0.4	2.0	39	—	—	—	3.3	630	−5.0
Tajikistan	5.8	2.0	3.3	32	—	—	—	4.4	360	−11.4
Uzbekistan	22.4	2.2	2.6	41	—	—	—	3.8	960	−2.3
Turkmenistan	4.4	4.6	2.8	45	—	—	—	3.9	—	—
Azerbaijan	7.5	1.0	1.8	56	—	—	—	2.5	500	−12.2
Georgia	5.4	−0.2	0.2	58	—	—	—	2.2	580^f	—
Iraq	20.3	2.8^c	3.5^c	71^c	—	—	—	5.7^d	—	—
Iran	62.5	2.4^c	2.9^c	57^e	—	—	—	—	—	—
Saudia Arabia	17.8	3.2	2.5	80	25.8	4,037	—	6.3	7,050	−1.7
Yemen	14.8	5.5	4.0	33	0.6	65	—	7.4	280	—
Oman	2.1	4.5	4.0	11^c	33.6	575	0.7	7.2^d	5,140	0.5
United Arab Emirates	2.4	2.9	1.8	83	90.2	1,478	—	4.1	—	0.4

Annexe table 1. *Continued*

	Population mid-1994 (millions)	Annual growth of population 1990–4 (%)	Annual growth of labour force (%)	Proportion urban 1994 (%)	Migrant stock 1990 (%)	Number of migrants 1990 (thousands)	Remittances as % of foreign exchange earnings	Total fertility rate 1994	GNP per capita 1994 ($US)	Annual average growth of GNP per capita 1985–94 (%)
Libya	5.2	3.5[c]	3.5[c]	70[c]	12.3	550	–	–	–	1.1
Kuwait	1.6	–6.8	–2.3	97	71.7	1,503	–	3.0	19,420	–2.4
Qatar	0.6	2.5	–	89[c]	63.5	299	–	4.3[d]	12,820	–
Brunei	0.3	2.1	–	58[c]	30.2	77	–	3.1[d]	14,240	–0.2
Sudan	27.3	2.7[c]	3.1[c]	–	3.3	803	25.1	–	–	–
Ethiopia	54.9	1.7	2.8	13	1.6	777	3.7	7.5	100	–2.3
Somalia	8.8	2.2[c]	1.5[c]	–	7.2	622	–	–	–	–
Kenya	26.0	2.7	3.4	27	0.7	168	1.2	4.9	250	–
Uganda	18.6	3.2	2.9	12	1.9	330	–	7.1	190	2.3
Tanzania	28.8	3.0	2.8	24	2.3	580	1.8	5.8	140	0.8
Madagascar	13.1	2.9	3.2	26	0.3	35	–	6.0	200	–1.7
Malawi	9.5	2.8	2.4	13	12.1	1,105	–	6.7	170	–0.7
Rwanda	7.8	2.6	2.9	6	1.0	69	2.3	–	80	–6.6
Burundi	6.2	3.0	2.9	7	6.1	333	–	6.7	160	–0.7
Zaire	42.5	3.2[c]	2.5[c]	39[c]	2.8	1,041	–	–	–	–1.0
Angola	10.4	3.6[c]	2.9[c]	28[c]	0.3	28	–	–	–	–6.8
Congo	2.6	3.1	2.7	58	5.9	129	–	6.7	620	–2.9
Central African Republic	3.2	2.5	2.1	39	2.0	57	–	5.7	370	–2.7
Gabon	1.3	3.2	1.9	49	8.9	100	–	5.5	3,880	–3.2
Chad	6.3	2.5	2.7	21	0.3	17	0.1	5.9	180	0.7
Niger	8.7	3.2	3.0	22	1.5	115	4.1	7.4	230	–2.1
Nigeria	108.0	2.9	2.8	38	0.3	254	5.3	5.6	280	1.2
Benin	5.3	2.9	2.7	41	1.0	48	17.8	6.1	370	–0.8
Togo	4.0	3.2	3.0	30	–	–	–	6.5	320	–2.7
Guinea	6.4	2.8	2.7	29	1.7	97	–	6.5	520	1.3
Côte d'Ivoire	13.8	3.6	2.9	43	29.3	3,440	–	6.5	610	–4.6
Ghana	16.6	2.8	3.0	36	0.9	137	1.7	5.3	410	1.4

Cameroon	13.0	3.0	2.9	44	2.4	267	1.3	5.7	680	-6.9
Mali	9.5	3.0	2.8	26	1.2	110	22.9	7.1	250	1.0
Burkina Faso	10.1	2.9	2.0	25	4.7	418	28.2	6.9	300	-0.1
Senegal	8.3	2.7	2.6	42	2.5	178	6.5	5.8	600	-0.7
Sierra Leone	4.4	2.4	2.3	35	5.0	198	–	6.5	160	-0.4
Gambia	1.1	3.9	3.4	25	11.2	101	–	5.4	330	0.5
Mauritania	2.2	2.5	2.7	52	3.3	65	0.5	5.2	480	0.2
St Kitts and Nevis	0.04	-0.3	–	49[e]	5.9	2	0.1	–	4,760	4.7
Antigua and Barbuda	0.07	0.6	–	32[e]	18.6	12	–	–	6,770	2.5
Dominica	–	-0.1	–	–	3.5	2	13.3	–	2,800	4.3
St Lucia	0.16	1.4	–	46[e]	4.0	5	0.1	–	3,130	4.0
St Vincent and the Grenadines	0.11	0.9	–	21[e]	5.0	61	1.2	–	2,140	4.5
Barbados	0.26	0.3	–	45[e]	10.5	27	–	1.8[d]	6,560	<0
Tonga	0.1	0.4	–	21[e]	6.4	6	38.7	–	1,590	0.3
Samoa	0.16	1.1	–	22[e]	3.6	6	–	4.5[d]	1,000	-0.3
Fiji	0.77	1.5	–	39[e]	1.7	12	2.9	3.0[d]	2,250	2.4
Kiribati	0.08	1.7	–	36[e]	3.5	3	15.7	–	740	–
Papua New Guinea	4.2	2.2	2.3	16	0.7	27	0.7	4.9	1,240	2.2
Solomon Islands	0.37	3.3	–	11	1.3	4	–	5.4[d]	810	2.2
Vanuatu	0.17	2.5	–	26	3.0	4	14.7	4.7[d]	1,150	-0.3
Micronesia (Federated States)	0.10	2.8	–	–	1.7	2	–	–	–	–
Maldives	0.25	3.3	–	29[e]	1.3	–	0.6	6.8	950	7.7

Sources: World Bank, *From plan to market: world development report 1996*, New York, Oxford University Press; United Nations Development Programme, *Human development report 1996*, New York, Oxford University Press; *International migration policies 1995* and editions of *World population* issued by United Nations, Department of Economic and Social Information and Policy Analysis, New York.

Notes: From the core extension, the allocation of countries to tiers is problematic as large countries such as China, India or Brazil transude several tiers. Countries or areas are allocated to the highest development tier in which their regions fall.
[a] 1980–93. [b] *Asia yearbook* estimates. [c] 1991–2000. [d] 1992. [e] 1993. [f] Mid-1990.

Annexe table 2. Immigration to the principal settler countries since the early 1980s

A. Canada: landed immigrants by country or region of last permanent residence, major regions and selected Asian countries, calendar years 1980–1993

Region/ country of last residence	1980	1981	1982	1983	1984	1985	1986	1987	1988	1989	1990	1991	1992	1993
All countries	143,117	128,618	121,147	89,157	88,239	84,302	99,219	152,098	161,929	192,001	214,230	230,781	252,842	255,819
Europe	41,168	46,295	46,150	24,312	20,901	18,859	22,709	37,563	40,689	52,105	51,945	44,055	44,871	46,602
Asia	71,602	48,830	41,617	36,906	41,896	38,597	41,600	67,327	81,136	93,213	111,739	119,955	139,216	147,323
Hong Kong	6,309	6,451	6,542	6,710	7,696	7,380	5,893	16,170	23,281	19,908	29,261	22,340	38,910	36,576
Japan	737	770	630	333	250	205	273	446	346	541	369	502	603	922
Singapore	290	389	435	241	176	166	220	489	1,141	1,634	1,077	807	616	592
South Korea	957	1,430	1,506	1,017	801	934	1,143	2,276	2,676	2,820	1,871	2,486	3,701	3,693
Taiwan	827	834	560	570	421	536	695	1,467	2,187	3,388	3,681	4,488	7,456	9,867
Africa	4,330	4,887	4,510	3,659	3,552	3,545	4,770	8,501	9,380	12,198	13,440	16,087	19,633	16,918
Oceania	2,497	2,251	2,119	1,213	1,151	1,128	1,227	1,826	1,822	2,041	2,647	3,135	3,659	3,082
North America	18,087	20,202	19,685	18,251	16,630	17,817	2,227	26,067	21,647	23,710	25,554	32,923	35,015	32,314
South America	5,433	6,136	6,870	4,816	4,084	4,356	6,686	10,801	7,255	8,685	8,898	10,582	10,389	9,580
Unknown/other	–	17	196	0	25		–	13	–	49	7	44	59	–

Source: Employment and Immigration Canada, *Annual immigration statistics*, Ottawa.

Annexe table 2. *Continued*

B. United States: immigrants by country or region of last permanent residence, major regions and selected Asian countries, fiscal years 1982–1994

Region/ country of last residence	1982	1983	1984	1985	1986	1987	1988	1989[a]	1990[a]	1991[a]	1992[a]	1993[a]	1994[a]
All countries	594,131	559,763	543,903	570,009	601,708	601,516	643,025	612,110	656,111	704,005	810,635	880,014	798,394
Europe	69,418	60,517	69,879	69,526	69,224	67,967	71,854	83,873	107,754	137,578	143,729	165,354	166,177
Asia	299,526	261,188	247,775	255,164	258,546	248,293	254,745	268,143	288,957	292,316	348,553	344,472	282,189
Hong Kong	11,908	12,525	12,290	10,795	9,930	8,785	11,817	12,236	12,858	15,564	16,741	14,010	11,949
Japan	4,084	4,234	4,517	4,552	4,444	4,711	5,085	5,454	5,690	5,212	11,676	7,648	6,969
Singapore	722	584	1,327	1,389	1,109	926	932	830	764	670	983	1,011	758
South Korea	30,697	31,449	32,537	34,791	35,164	35,397	34,151	31,604	28,720	20,808	18,374	17,245	15,395
Taiwan	12,099	19,018	14,684	17,517	15,931	14,080	12,376	14,705	16,344	15,067	17,905	15,736	11,157
Africa	11,966	4,386	13,594	15,263	15,500	15,730	17,124	16,427	17,475	20,643	24,826	25,179	24,768
Oceania	4,046	3,510	4,249	4,552	4,352	4,473	4,324	4,450	5,055	4,904	4,485	6,050	5,632
North America	154,847	164,064	169,475	185,467	211,428	220,244	253,260	193,869	189,328	200,694	238,552	285,758	272,305
South America	32,803	35,169	38,636	40,052	42,650	44,782	41,646	45,275	47,327	47,771	50,488	53,201	47,321
Unknown/other	21,525	22,874	295	12	8	63	72	73	215	99	2	4	4

Source: US Department of Justice, *Statistical yearbooks* of the Immigration and Naturalization Service, Washington, DC.
  [a] Excludes IRCA legalization. When this is included, the total intake was 1,090,924 in 1989, 1,536,483 in 1990, 1,827,167 in 1991, 973,977 in 1992, 904,292 in 1993 and 804,416 in 1994. The largest number legalized under this programme came from Mexico, Central and South America, and the Caribbean.

Annexe table 2. *Continued*

C. Australia: settler arrivals by country or region of last permanent residence, major regions and selected Asian countries, financial years 1982/3–1993/4

Region/ country of last residence	1982/3	1983/4	1984/5	1985/6	1986/7	1987/8	1988/9	1989/90	1990/1	1991/2	1992/3	1993/4
All countries	93,177	69,808	78,087	92,410	113,309	143,490	145,316	121,227	121,688	107,391	76,330	69,768
Europe	46,980	24,470	22,134	27,144	36,119	42,879	41,367	37,820	32,352	26,847	22,152	20,013
Asia	26,450	28,504	33,689	34,537	42,098	54,578	58,471	53,588	65,664	59,057	36,254	30,444
Hong Kong	2,756	3,691	5,136	4,912	5,140	7,942	9,998	11,538	16,747	15,656	8,111	4,075
Japan	331	327	286	358	479	873	840	710	634	577	500	446
Singapore	1,044	944	1,142	1,215	2,222	2,817	2,806	2,137	1,944	1,652	894	779
South Korea	569	564	636	1,201	1,510	1,756	1,627	1,338	948	1,174	885	650
Taiwan	176	233	344	575	987	1,320	1,884	2,889	3,261	2,946	1,389	779
Africa	4,937	3,604	3,337	5,816	8,528	8,495	5,694	4,403	4,393	3,671	3,207	4,019
Oceania	9,385	8,094	12,072	17,296	18,571	28,832	31,581	17,367	11,818	11,156	10,554	11,281
North America	3,802	3,079	3,174	3,566	3,769	4,102	3,974	3,887	3,645	3,347	2,539	2,538
South America	1,623	2,057	3,681	4,051	4,224	4,580	4,192	4,090	3,650	3,213	1,483	1,096
Not stated	0	0	0	0	0	24	37	72	166	100	141	377

Source: Bureau of Immigration and Population Research, *Australian immigration: consolidated statistics*, No. 17 (1991–92), Canberra; Bureau of Immigration, Multicultural and Population Research, *Australian immigration: consolidated statistics*, No. 18 (1993–4), Canberra.

References

Abella M I 1992 International migration and development. In Battistella and Paganoni 1992: 22–40

Abella M I 1994a Introduction [to 'Turning points in labor migration']. *Asian and Pacific Migration Journal* **3**(1): 1–6

Abella M I (ed.) 1994b Turning points in labor migration. *Asian and Pacific Migration Journal* **3**(1): special issue

Abu-Lughod J L 1995 The displacement of the Palestinians. In Cohen 1995a: 410–13

Ackland R, Williams L 1992 *Immigrants and the Australian labour market: the experience of three recessions*. Canberra, Australian Government Publishing Service

Adams P G 1983 *Travel literature and the evolution of the novel*. Lexington, University Press of Kentucky

Adas M 1989 *Machines as the measure of men: science, technology and ideologies of western dominance*. Ithaca, Cornell University Press

Addleton J S 1992 *Undermining the centre: the Gulf migration and Pakistan*. Karachi, Oxford University Press

Adelman H 1995 The Palestinian diaspora. In Cohen 1995a: 414–17

Adelman J 1995 European migration to Argentina. In Cohen 1995a: 215–19

Adepoju A 1991 Binational communities and labor circulation in sub-Saharan Africa. In Papademetriou and Martin 1991: 45–64

Adepoju A 1995a Emigration dynamics in sub-Saharan Africa. *International Migration* **33**(3/4): 315–90

Adepoju A 1995b Migration in Africa: an overview. In Baker and Aina 1995: 87–108

Adomako-Sarfoh J 1974 The effects of expulsion of migration workers on Ghana's economy, with particular reference to the cocoa industry. In S Amin and D Forde (eds) *Modern migrations in western Africa*. Oxford, Oxford University Press: 138–55

Ahmad A 1992 *In theory: classes, nations, literatures*. London, Verso

Ainsworth M, Over M 1994 AIDS and development. *The World Bank Research Observer* **9**(2): 203–40

Aktar C, Ogelman N 1994 Recent developments in East–West migration: Turkey and the petty traders. *International Migration* **32**(2): 343–54

Allsop K 1967 *Hard travellin': the story of the migrant worker*. London, Hodder and Stoughton

Almeida L D 1992 And they still haven't found what they're looking for: a survey of the New Irish in New York city. In O'Sullivan 1992: 196–221

Amin S 1995 Migrations in contemporary Africa: a retrospective view. In Baker and Aina 1995: 29–40

218   *Migration and Development*

Amjad R 1996 Philippines and Indonesia: on the way to a migration transition? *Asian and Pacific Migration Journal* 5(2–3): 339–66

AMWC 1991 *Foreign domestic workers in Hong Kong: a baseline study.* Hong Kong, Asian Migrant Workers' Centre

Anderson B 1983 *Imagined communities: reflections on the origin and spread of nationalism.* London, Verso

Appleyard R T (ed.) 1989 *The impact of international migration on developing countries.* Paris, OECD

Archavanitkul K, Guest P 1994 Migration and the commercial sex sector in Thailand. *Health Transition Review*, supplement to vol. 4: 273–95

Arcinas F R 1986 The Philippines. In Gunatilleke 1986: 259–305

Arlacchi P 1983 *Mafia, peasants and great estates: society in traditional Calabria.* Cambridge, Cambridge University Press

Bade K J 1995 Germany: migrations in Europe up to the end of the Weimar Republic. In Cohen 1995a: 131–5

Bailyn B 1987 *Voyagers to the west: a passage in the peopling of America on the eve of the Revolution.* New York, Knopf

Baines D 1986 *Migration in a mature economy: emigration and internal migration in England and Wales 1861–1900.* Cambridge, Cambridge University Press

Baines D 1991 *Emigration from Europe 1815–1930.* London, Macmillan

Baines D 1994 European labor markets, emigration and internal migration 1850–1913. In T J Hatton and J G Williamson (eds) *Migration and the international labor market 1850–1939.* New York, Routledge: 35–54

Bairoch P 1986 Historical roots of economic underdevelopment: myths and realities. In W J Mommsen and J Osterhammel (eds) *Imperialism and after: continuities and discontinuities.* London, Allen and Unwin: 191–216

Bairoch P 1988 *Cities and economic development: from the dawn of history to the present.* Chicago, University of Chicago Press

Baker H D R 1994 Branches all over: the Hong Kong Chinese in the United Kingdom. In Skeldon 1994: 291–307

Baker J, Aina T A (eds) 1995 *The migration experience in Africa.* Uppsala, Nordiska Afrikainstitutet

Ball B, Demko G J 1978 Internal migration in the Soviet Union. *Economic Geography* 54(2): 95–114

Banister J 1987 *China's changing population.* Stanford, Stanford University Press

Barff R, Ellis M, Reibel M 1995 The links between immigration and internal migration in the United States: a comparison of the 1970s and 1980s. Working Paper Series No. 1, The Nelson A Rockefeller Centre for the Social Sciences, Dartmouth College

Barnett T, Blaikie P 1992 *AIDS in Africa: its present and future impact.* London, Belhaven

Battistella G, Paganoni A (eds) 1992 *Philippine labor migration: impact and policy.* Quezon City, the Philippines, Scalabrini Migration Center

Baydar N, White M J, Simkins C, Babakol O 1990 Effects of agricultural development policies on migration in Peninsular Malaysia. *Demography* 27(1): 97–109

Beaverstock J V 1994 Rethinking skilled international labour migration: world cities and banking organizations. *Geoforum* 25(3): 323–8

Bedford R D 1992 International migration in the South Pacific region. In Kritz, Lim and Zlotnik 1992: 41–62

Berry B J L 1993 Transnational urbanward migration, 1830–1980. *Annals of the Association of American Geographers* 83(3): 389–405

Bertram I G, Watters R F 1985 The MIRAB economy in South Pacific microstates. *Pacific Viewpoint* 26(3): 497–519

Bertram I G, Watters R F 1986 The MIRAB process: earlier analyses in context. *Pacific Viewpoint* 27(1): 47–59

Bilsborrow R F 1992 Rural poverty, migration, and the environment in developing countries: three case studies. Working paper WPS1017, Washington DC, World Bank

Birrell R, Dobson I 1994 Austudy dependence amongst students in Australia. *People and Place* 2(4): 30–7

Black R 1995 Review of S Spencer, *Strangers and citizens: a positive approach to migrants and refugees. Transactions Institute of British Geographers* NS 20(2): 267–9

Blaut J M 1993 *The colonizer's model of the world: geographical diffusionism and Eurocentric history.* New York, The Guilford Press

Boehmer E 1995 *Colonial and postcolonial literature.* Oxford, Oxford University Press

Bogue D J 1969 *Principles of demography.* New York, Wiley

Böhning W R 1995 Undesired jobs and what one can do to fill them: the case of the Republic of Korea. In M I Abella, Y-B Park and W R Böhning *International migration papers 1. Adjustments to labour shortages and foreign workers in the Republic of Korea.* Geneva, International Labour Office: 19–35

Böhning W R, Schloeter-Paredes M-L (eds) 1994 *Aid in place of migration?* Geneva, International Labour Office

Bonacich E 1993 The other side of ethnic entrepreneurship: a dialogue with Waldinger, Aldrich, Ward and associates and reply to Waldinger. *International Migration Review* 27(3): 685–92; 701–2

Bongaarts J 1996 Global trends in AIDS mortality. *Population and Development Review* 22(1): 21–45

Borjas G J 1990 *Friends or strangers: the impact of immigrants on the U.S. economy.* New York, Basic Books

Borjas G J 1994 The economics of immigration. *Journal of Economic Literature* 32(4):1667–1717

Borjas G J 1995 The economic benefits from immigration. *Journal of Economic Perspectives* 9(2): 3–22

Boserup E 1970 *Women's role in economic development.* London, Allen and Unwin

Bowman I 1916 *The Andes of southern Peru.* New York, The American Geographical Society

Brimelow P 1995 *Alien nation: common sense about America's immigration disaster.* New York, Random House

Brookfield H 1974 *Interdependent development.* London, Methuen

Brown L A 1990 *Place, migration and development in the Third World: an alternative view.* London, Routledge

Brown R P C 1994 Migrants' remittances, savings and investment in the South Pacific. *International Labour Review* **133**(3): 347–67

Brown R P C, Connell J 1993 The global flea market: migration, remittances and the informal economy in Tonga. *Development and Change* **24**(4): 611–47

Brown R P C, Connell J (eds) 1995 Migration and remittances in the South Pacific. *Asian and Pacific Migration Journal* **4**(1): special issue

Brydon L 1992 Ghanaian women in the migration process. In Chant 1992: 91–108

Burki S J 1991 Migration from Pakistan to the Middle East. In Papademetriou and Martin 1991: 139–61

Cahill D 1990 *Intermarriages in international contexts.* Quezon City, the Philippines, Scalabrini Migration Center

Calavita K 1994 U.S. immigration and policy responses: the limits of legislation. In Cornelius, Martin and Hollifield 1994: 55–82

Caldwell J C 1969 *African rural–urban migration: the movement to Ghana's towns.* New York, Columbia University Press

Campbell D 1994 Foreign investment, labour immobility and the quality of employment. *International Labour Review* **133**(2): 185–204

Cariño B V 1992 Migrant workers from the Philippines. In Battistella and Paganoni 1992: 4–21

Castles S 1995 Contract labour migration. In Cohen 1995a: 510–14

Castles S, Miller M J 1993 *The age of migration: international population movements in the modern world.* London, Macmillan

Chaliand G, Rageau J-P 1995 *The Penguin atlas of diasporas.* New York, Viking

Champion A G 1992 Urban and regional demographic trends in the developed world. *Urban Studies* **39**(3/4): 461–82

Champion A G 1994 Population change and migration in Britain since 1981: evidence for continuing deconcentration. *Environment and Planning A* **26**(10): 1501–20

Champion A G 1995 The counterurbanization cascade: an analysis of the 1991 census special migration statistics for Great Britain. Paper presented at the International Conference on Population Geography, Dundee, 16–19 September

Chan K W 1994 Urbanization and rural–urban migration in China since 1982: a new baseline. *Modern China* **20**(3): 243–81

Chan K W, Xu X 1985 Urban population growth and urbanization in China since 1949: reconstructing a baseline. *China Quarterly* **104**: 583–613

Chan S 1990 European and Asian immigration into the United States in comparative perspective, 1820s to 1920s. In V Yans-McLaughlin (ed.) *Immigration reconsidered: history, sociology and politics.* New York, Oxford University Press: 37–75

Chant S (ed.) 1992 *Gender and migration in developing countries.* London, Belhaven

Chapman M (ed.) 1985 Mobility and identity in the island Pacific. *Pacific Viewpoint* **26**(1): special issue

Chapman M, Prothero R M (eds) 1985 *Circulation in population movement: substance and concepts from the Melanesian case*. London, Routledge and Kegan Paul

Chase R S, Hill E B, Kennedy P 1996 Pivotal states and U.S. strategy. *Foreign Affairs* **75**(1): 33–51

Chase-Dunn C 1985 The system of world cities, AD 800–1975. In M Timberlake (ed.) *Urbanization in the world economy*. New York, Academic Press: 269–92

Chatelain A 1976 *Les migrants temporaires en France de 1800 à 1914*. Lille, Presses Universitaires de Lille

Chen C, Liu S-F 1996 Migration into and out of Taiwan 1895–1944. Paper presented at the Conference on Asian Population History, IUSSP and the Institute of Economics, Academia Sinica, Taipei, 4–8 January

Cheng L-K 1985 *Social change and the Chinese in Singapore*. Singapore, Singapore University Press

Chesnais J-C 1992 *The demographic transition: stages, patterns and economic implications. A longitudinal survey of sixty-seven countries covering the period 1720–1984*. Oxford, Clarendon Press

Chiengkul W 1986 Thailand. In Gunatilleke 1986: 306–37

Chiew S K 1995 Citizens and foreign labour in Singapore. In J H Ong, K B Chan and S B Chew (eds) *Crossing borders: transmigration in Asia Pacific*. Singapore, Prentice Hall: 472–86

Chung J S 1995 Economy: the Democratic People's Republic of Korea. In *The Far East and Australasia, 1995*, 26th edn. London, Europa Publications: 459–65

Clarke C, Peach C, Vertovec S 1990 Introduction: themes in the study of the South Asian diaspora. In C Clarke, C Peach and S Vertovec (eds) *South Asians overseas: migration and ethnicity*. Cambridge, Cambridge University Press: 1–29

Cleland J, Wilson C 1987 Demand theories of the fertility transition: an iconoclastic view. *Population Studies* **41**(1): 5–30

Clifford J 1992 Traveling cultures. In L Grossberg, C Nelson and P Treichler (eds) *Cultural studies*. New York, Routledge: 96–116

Clout H, Salt J 1976 The demographic background. In J Salt and H Clout (eds) *Migration in post-war Europe: geographical essays*. London, Oxford University Press: 7–29

Cohen A 1971 Cultural strategies in the organization of trading diasporas. In C Meillassoux (ed.) *The development of indigenous trade and markets in West Africa*. International African Institute, Oxford University Press: 266–81

Cohen R 1981 The new international division of labour: multinational corporations and the urban hierarchy. In M Dear and A J Scott (eds) *Urbanization and urban planning in capitalist society*. London, Methuen: 287–315

Cohen R 1987 *The new helots: migrants in the international division of labour*. Aldershot, Avebury

Cohen R (ed.) 1995a *The Cambridge survey of world migration*. Cambridge, Cambridge University Press

Cohen R 1995b Prologue. In Cohen 1995a: 1–9

Cohen R 1996a Diasporas and the nation state: from victims to challengers. *International Affairs* **72**(3): 507–20

Cohen R (series ed.) 1996b *International library of studies on migration*, 6 vols. Cheltenham, Edward Elgar

Coleman D A 1992 Does Europe need immigrants? Population and work force projections. *International Migration Review* **26**(2): 413–61

Collier P 1995 The marginalization of Africa. *International Labour Review* **134**(4–5): 541–57

Compton P 1976 Migration in eastern Europe. In Salt and Clout 1976: 168–215

Connell J 1983–5 *Migration, employment and development in the South Pacific.* Country reports, 22 vols. Noumea, International Labour Organization and South Pacific Commission

Connell J (ed.) 1990a *Migration and development in the South Pacific.* Canberra, National Centre for Development Studies, Australian National University

Connell J 1990b Modernity and its discontents: migration and change in the South Pacific. In Connell 1990a: 1–28

Connell J 1994 Beyond the reef: migration and agriculture in Micronesia. *ISLA: A Journal of Micronesian Studies* **2**(1): 83–101

Conway D 1994 The complexity of Caribbean migration. *Caribbean Affairs* **7**(4): 96–119

Copper J F 1990 *Taiwan: nation state or province?* Boulder, Westview Press

Cornelius W A 1994 Japan: the illusion of immigration control. In Cornelius, Martin and Hollifield 1994: 375–410

Cornelius W A, Martin P L, Hollifield J F (eds) 1994 *Controlling immigration: a global perspective.* Stanford, Stanford University Press

Courgeau D, Lelièvre E 1992 *Event analysis in demography.* Oxford, Clarendon Press

Coussy J 1994 The outlook for economic integration in sub-Saharan Africa: effects on continental and intercontinental migration. In *Migration and development: new partnerships for cooperation.* Paris, OECD: 240–9

Cowen M P, Shenton R W 1996 *Doctrines of development.* London, Routledge

Cresswell T 1993 Mobility as resistance: a geographical reading of Kerouac's 'On the road'. *Transactions Institute of British Geographers* **18**(2): 249–62

Crush J, Jeeves A, Yudelman D 1991 *South Africa's labor empire: a history of black migrancy to the gold mines.* Boulder, Westview Press

Curtin P D 1984 *Cross-cultural trade in world history.* Cambridge, Cambridge University Press

Curtin P D 1989 *Death by migration: Europe's encounter with the tropical world in the nineteenth century.* Cambridge, Cambridge University Press

Davis K 1963 The theory of change and response in modern demographic history. *Population Index* **29**(4): 345–66

Decosas J, Kane F, Anarfi J K, Sodji K D R, Wagner H U 1995 Migration and AIDS. *The Lancet* **346**(8978): 826–8

Denevan W (ed.) 1992 *The native populations of the Americas in 1492.* Madison, University of Wisconsin Press

De Soto H 1989 *The other path: the invisible revolution in the Third World.* New York, Harper and Row

Devine T M 1979 Temporary migration and the Scottish highlands in the nineteenth century. *Economic History Review* **32**(3): 344–59

Díaz-Briquets S 1991 The effects of international migration on Latin America. In Papademetriou and Martin 1991: 183–99

Dicken P 1992 *Global shift: the internationalization of economic activity,* 2nd edn. London, Paul Chapman

Douglas I, Gu H, He M 1994 Water resources and environmental problems of China's great rivers. In D Dwyer (ed.) *China: the next decades.* London, Longman

Dunlop J B 1993 Will a large-scale migration of Russians to the Russian Republic take place over the current decade? *International Migration Review* **27**(3): 605–29

Edwards C 1985 *The fragmented world: competing perspectives on trade, money and crisis.* London, Methuen

Ee J 1961 Chinese migration to Singapore 1896–1941. *Journal of the Southeast Asian Historical Society* **2**(1): 42–62

Elizaga J C 1970 *Migraciones a las areas metropolitanas de América Latina.* Santiago de Chile, Centro Latinoamericano de Demografía

ESCAP 1975 *Population of the Republic of Korea.* Country Monograph Series No. 2. Bangkok, United Nations Economic and Social Commission for Asia and the Pacific

ESCAP 1980 *Migration, urbanization and development in Sri Lanka.* Bangkok, United Nations Economic and Social Commission for Asia and the Pacific

Fahim Khan M 1986 Pakistan. In Gunatilleke 1986: 110–65

Fang Shan 1991 Mainland China's overseas construction contracts and export of labour. *Issues and Studies* **27**(2): 65–75

Farris W W 1985 *Population, disease and land in early Japan, 645–900.* Cambridge, Mass., Harvard University Press

Fernandez-Armesto F 1995 *Millennium: a history of our last thousand years.* London, Bantam Press

Ferris E G 1993 *Beyond borders: refugees, migrants and human rights in the post-cold war era.* Geneva, WCC Publications

Fields G S 1994 The migration transition in Asia. *Asian and Pacific Migration Journal* **3**(1): 7–30

Findlay A M 1990 A migration channels approach to the study of high level manpower movements. *International Migration* **28**(1): 15–22

Findlay A M 1991 Third World migration: divergent views on the way ahead. *Newsletter* of the Developing Areas Research Group No. 15

Findlay A M 1994 *The Arab world.* London, Routledge

Findlay A M 1995 Skilled transients: the invisible phenomenon? In Cohen 1995a: 512–22

Findlay A, Findlay A 1982 *The geographical interpretation of international migration: a case study of the Maghreb.* University of Durham, Centre for Middle Eastern and Islamic Studies, Occasional Papers Series No. 14

Findley S, Traoré S, Ouedraogo D, Diarra S 1995 Emigration from the Sahel. *International Migration* **33**(3/4): 469–520

Fischer D H 1989 *Albion's seed: four British folkways in America*. New York, Oxford University Press

Francks P 1992 *Japanese economic development: theory and practice*. London, Routledge

Frey W H 1993 The new urban revival in the United States. *Urban Studies* **30**(4/5): 741–74

Frey W H 1995 Immigration and internal migration 'flight' from US metropolitan areas: toward a new demographic balkanization. *Urban Studies* **32**(4/5): 733–57

Friedberg R M, Hunt J 1995 The impact of immigrants on host country wages, employment and growth. *Journal of Economic Perspectives* **9**(2): 23–44

Friedmann J, Wulff R 1975 *The urban transition: comparative studies of newly industrializing societies*. London, Edward Arnold

Fry R 1996 Has the quality of immigrants declined? Evidence from the labor market attachment of immigrants and natives. *Contemporary Economic Policy* **14**(3): 53–70

Fuchs R J, Demko G J 1978 The postwar mobility transition in Eastern Europe. *Geographical Review* **68**(2): 171–82

Gall S B, Gall T L (eds) 1993 *Statistical record of Asian Americans*. Detroit, Gale Research Inc.

Garrett L 1996 The return of infectious disease. *Foreign Affairs* **75**(1): 66–79

Garson J-P 1992 Migration and interdependence: the migration system between France and Africa. In Kritz, Lim and Zlotnik 1992: 80–93

Gellner E 1992 *Postmodernism, reason and religion*. London, Routledge

Giddens A 1981 *A contemporary critique of historical materialism*. Vol. 1: *Power, property and the state*. London, Macmillan

Ginsburg N 1990 *The urban transition: reflections on the American and Asian experiences*. Hong Kong, The Chinese University Press

Ginsburg N 1991 Extended metropolitan regions in Asia: a new spatial paradigm. In N Ginsburg, B Koppel and T G McGee (eds) *The extended metropolis: settlement transition in Asia*. Honolulu, University of Hawaii Press: 27–46

Gitmez A S 1991 Migration without development: the case of Turkey. In Papademetriou and Martin 1991: 115–34

Gogate S 1986 India. In M I Abella and Y Atal (eds) *Middle East interlude: Asian workers abroad. A comparative study of four countries*. Bangkok, UNESCO Regional Office: 27–105

Golini A, Bonifazi C, Righi A 1993 A general framework for the European migration system in the 1990s. In King 1993a: 67–82

Gottschang T R 1982 *Migration from north China to Manchuria: an economic history 1891–1942*. Ann Arbor, University of Michigan

Gould W T S 1995 Migration and recent economic and environmental change in East Africa. In Baker and Aina 1995: 122–45

Gould W T S, Prothero R M 1975 Space and time in African population mobility. In L A Kosiński and R M Prothero (eds) *People on the move: studies on internal migration*. London, Methuen: 39–49

Grecic V 1993 Mass migration from Eastern Europe: a challenge to the West? In King 1993a: 135–51

Green A E 1995 The geography of dual career households: a research agenda and selected evidence from secondary data sources for Britain. *International Journal of Population Geography* **1**(1): 51–67

Grigg D B 1977 E G Ravenstein and the 'laws of migration'. *Journal of Historical Geography* **3**(1): 41–54

Gunatilleke G (ed.) 1986 *Migration of Asian workers to the Arab world.* Tokyo, United Nations University

Gunatilleke G (ed.) 1992 *The impact of labour migration on households: a comparative study of seven Asian countries.* Tokyo, United Nations University

Gwynne R N 1987 Modern manufacturing growth in Latin America. In D Preston (ed.) *Latin American development: geographical perspectives.* London, Longman: 102–40

Hackett P, Summerscale J 1995 Venezuela: economy. In *South America, Central America and the Caribbean 1995*, 5th edn. London, Europa Publications: 637–42

Halfacree K, Boyle P 1993 The challenge facing migration research: the case for a biographical approach. *Progress in Human Geography* **17**(3): 333–48

Hambro E 1955 *The problem of Chinese refugees in Hong Kong.* Leiden, A W Sijthoff

Handlin O 1951 *The uprooted.* Cambridge, Mass., Harvard University Press

Hanley S B 1973 Migration and economic change in Okayama during the Tokugawa period. *Keio Economic Studies* **10**(2): 19–35

Hanlon G 1992 Graduate emigration: a continuation or a break with the past? In O'Sullivan 1992: 183–95

Harrell P 1992 *Sowing the seeds of change: Chinese students, Japanese teachers.* Stanford, Stanford University Press

Harris C D 1982 The urban and industrial transformation of Japan. *The Geographical Review* **72**(1): 50–89

Harris N 1986 *The end of the Third World: newly industrializing countries and the decline of an ideology.* Harmondsworth, Penguin Books

Harris N 1995 Stopping immigration. *Development and Policy Review* **13**(1): 85–9

Hayami A 1973 Labour migration in a pre-industrial society: a study tracing the life histories of the inhabitants of a village. *Keio Economic Studies* **10**(2): 1–17

Hayes G 1992 Polynesian migration and the demographic transition: a missing dimension of recent theoretical models. *Pacific Viewpoint* **33**(1): 1–35

Helweg A W 1991 Indians in Australia: theory and methodology of the new immigration. In S Vertovec (ed.) *Aspects of the South Asian diaspora. Oxford University Papers on India*, Vol. 2, Part 2. Delhi, Oxford University Press: 7–35

Herzog L A 1991 Cross-national urban structure in the era of global cities: the US–Mexico transfrontier metropolis. *Urban Studies* **28**(4): 519–33

Hettne B 1995 *Development theory and the three worlds*, 2nd edn. London, Longman

Ho S P S 1978 *Economic development of Taiwan.* New Haven, Yale University Press

Hobsbawm E 1994 *Age of extremes: the short twentieth century 1914–1991*. London, Michael Joseph

Holborn L 1968 Refugee migration in the twentieth century. In F D Scott (ed.) *World migration in modern times*. Englewood Cliffs, NJ, Prentice Hall: 153–9

Hondagneu-Sotelo P 1994 *Gendered transitions: Mexican experiences of immigration*. Berkeley, University of California Press

Hong Kong 1992 *1992 survey of overseas investment in Hong Kong's manufacturing industries*. Hong Kong Government, Industry Department

Hopkins K (ed.) 1971 *Hong Kong: the industrial colony*. Hong Kong, Oxford University Press

Howe C 1996 *The origins of Japanese trade supremacy*. Hong Kong, Oxford University Press

Hsieh, K C, Tsai H T, Yu S Y, Yeh Y Y, Cheng Y C 1989 *An investigation of the current situation and problems of emigration* (in Chinese). Taipei, Development Review Council

Hsieh Y-W 1985 Urban deconcentration in developing countries: an analysis of the processes of population dispersion in Taiwan and South Korea. Doctoral dissertation, Department of Urban Planning and Policy Development, Rutgers University. Ann Arbor, University Microfilms

Hugo G 1993 Indonesian labour migration to Malaysia: trends and policy implications. *Southeast Asia Journal of Social Science* 21(1): 36–7

Hugo G 1994a *The economic implications of emigration from Australia*. Canberra, Bureau of Immigration and Population Research, Australian Government Publishing Service

Hugo G 1994b Migration as a survival strategy: the family dimension of migration. In *Population distribution and migration*. Proceedings of the United Nations Expert Meeting on Population Distribution and Migration, Santa Cruz, Bolivia, 18–22 January 1993. New York: 168–82

Hui W T 1992 Singapore's immigration policy: an economic perspective. In L Low and N H Yoh (eds) *Public policies in Singapore: changes in the 1980s and future signposts*. Singapore, Times Academic Press: 161–84

Hunt D 1989 *Economic theories of development: an analysis of competing paradigms*. New York, Harvester Wheatsheaf

Huntington S P 1993 The clash of civilizations? *Foreign Affairs* 72(3): 22–49

Iguchi Y 1996 Japan. Paper presented at the meeting on International Migration and the Labour Market in Asia: National Policies and Regional Cooperation, Tokyo, Japan Institute of Labour, 1–2 February

IIE 1994 *Open doors 1993–1994: report on international educational exchange*. New York, Institute of International Education

INS 1994 *Statistical yearbook of the Immigration and Naturalization Service, 1993*. Washington DC, US Government Printing Office

Jackson R T 1989 Filipino migration to Australia: the image and a geographer's dissent. *Australian Geographical Studies* 27(2): 170–81

Japan 1996 *Japan statistical yearbook 1995*. Tokyo, Statistics Bureau, Management and Coordination Agency

Jones H 1990 *Population geography*, 2nd edn. London, Paul Chapman

Josey A 1971 *Lee Kuan Yew*. Singapore, Donald Moore

Kang S-D 1995 Data on international migration to Korea. *Asian and Pacific Migration Journal* 4(4): 579–84

Katzen L 1995 South Africa: economy. In *Africa south of the Sahara 1995*, 24th edn. London, Europa Publications: 850–6

Kawabe H 1984 Internal migration. In *Population of Japan*. Bangkok, United Nations Economic and Social Commission for Asia and the Pacific, Country Monograph No. 11: 123–39

Kee P, Skeldon R 1994 The migration and settlement of Hong Kong Chinese in Australia. In Skeldon 1994: 183–96

Kelley A C 1991 The Human Development Index: 'handle with care'. *Population and Development Review* 17(2): 315–24

Kim D-S 1994 The demographic transition in the Korean peninsula, 1910–1990: South and North Korea compared. *Korea Journal of Population and Development* 23(2): 131–55

Kim D Y, Sloboda J E 1981 Migration and Korean development. In R Repetto, T H Kwon, S U Kim and O Y Kim (eds) *Economic development, population policy and demographic transition in the Republic of Korea*. Cambridge, Mass., Harvard University Press: 36–138

King A D 1990 *Global cities: post-imperialism and the internationalization of London*. London, Routledge

King R (ed.) 1993a *The new geography of European migrations*. London, Belhaven

King R 1993b Why do people migrate? The geography of departure. In King 1993a: 17–46

King R 1996 Migration in a world historical perspective. In van den Broek 1996: 7–75

King R, Connell J, White P (eds) 1995 *Writing across worlds: literature and migration*. London, Routledge

King R, Rybaczuk K 1993 Southern Europe and the international division of labour: from emigration to immigration. In King 1993a: 175–206

Kirkby R J R 1985 *Urbanization in China: town and country in a developing economy 1949–2000 AD*. New York, Columbia University Press

Kleiner R J, Sørensen T, Dalgard O D, Moum T, Drews D 1986 International migration and internal migration: a comprehensive theoretical approach. In I A Glazier and L de Rosa (eds) *Migration across time and nations: population mobility in historical contexts*. New York, Holmes and Meier: 305–17

Knight J 1994 The temple, the town office and the migrant: demographic pluralism in rural Japan. *European Journal of Sociology* 35(1): 21–47

Kojima R 1995 Urbanization in China. *The Developing Economies* 33(2): 121–54

Korcelli P 1992 International migrations in Europe: Polish perspectives for the 1990s. *International Migration Review* 26(2): 292–304

Kotkin J 1993 *Tribes: how race, religion and identity determine success in the new global economy*. New York, Random House

Kritz M, Lim L L, Zlotnik H (eds) 1992 *International migration systems: a global approach.* Oxford, Clarendon Press

Kumar S 1994 Johor–Singapore–Riau growth triangle: a model of subregional cooperation. In Thant, Tang and Kakazu 1994: 175–217

Kunieda M 1996 Foreign worker policy and illegal migration in Japan. In *Migration and the labour market in Asia: prospects to the year 2000.* Paris, OECD: 195–206

Kunz E F 1981 Part II: the analytic framework. Exile and resettlement: refugee theory. *International Migration Review* 15(1–2): 42–51

Kuroda T 1986 Urbanization of the population and development: new tasks. In *Urbanization and development in Japan.* Population and Development Series No. 3. Tokyo, Asian Population and Development Association

Kuznets P W 1987 Koreans in America: recent migration from South Korea to the United States. In S Klein (ed.) *The economics of mass migration in the twentieth century.* New York, Paragon: 41–69

Lam L 1994 Searching for a safe haven: the migration and settlement of Hong Kong Chinese immigrants in Toronto. In Skeldon 1994: 163–79

Langton J, Hoppe G 1990 Urbanization, social structure and population circulation in pre-industrial times: flows of people through Vadstena (Sweden) in the mid-nineteenth century. In P J Corfield and D Keene (eds) *Work in towns 850–1850.* Leicester, Leicester University Press: 138–63

Ledent J 1982 The factors of urban population growth: net immigration versus natural increase. *International Regional Science Review* 7(2): 99–125

Lee C, de Vos G (eds) 1981 *Koreans in Japan: ethnic conflict and accommodation.* Berkeley, University of California Press

Lee E S 1966 A theory of migration. *Demography* 3(1): 47–57

Lemon A 1980 Migrant labour in western Europe and southern Africa. In A Lemon and N Pollock (eds) *Studies in overseas settlement and population.* London, Longman: 127–58

Lemon A 1982 Migrant labour and frontier commuters: reorganizing South Africa's black labour supply. In D M Smith (ed.) *Living under apartheid: aspects of urbanization and social change in South Africa.* London, Allen and Unwin: 64–89

Lin C-P 1994 China's students abroad: rates of return. *The American Enterprise* 5(6): 12–14

Little T 1995 Israel: history. In *The Middle East and North Africa 1995,* 41st edn. London, Europa Publications: 511–30

Long L 1988 *Migration and residential mobility in the United States.* New York, Russell Sage Foundation

Long L 1991 Residential mobility differences among developed countries. *International Regional Science Review* 14(2): 133–47

Lovejoy P E 1989 The impact of the Atlantic slave trade on Africa: a review of the literature. *Journal of African History* 30: 365–94

Lozano Ascencio F 1993 *Bringing it back home: remittances to Mexico from migrant workers in the United States.* San Diego, Center for

US–Mexican Studies, University of California at San Diego

Mabogunje A L 1970 Systems approach to a theory of rural–urban migration. *Geographical Analysis* **2**(1): 1–17

McDowell L 1996 Off the road: alternative views of rebellion, resistance and the 'beats'. *Transactions Institute of British Geographers* **21**(2): 412–19

McGee T G 1991 The emergence of *desakota* regions in Asia: expanding a hypothesis. In N Ginsburg, B Koppel and T G McGee (eds) *The extended metropolis: settlement transition in Asia*. Honolulu, University of Hawaii Press: 3–25

McGee T G 1994 The future of urbanisation in developing countries: the case of Indonesia. *Third World Planning Review* **16**(1): iii–xii

MacLaughlin J 1994 *Ireland: the emigrant nursery and the world economy*. Cork, Cork University Press

MacMaster N 1995 Labour migration in French North Africa. In Cohen 1995a: 190–5

McNeill W H 1986 *Polyethnicity and national unity in world history*. Toronto, University of Toronto Press

Makinwa-Adebusoye P K 1995 Emigration dynamics in west Africa. *International Migration* **33**(3/4): 435–67

Mandel M S 1995 One nation indivisible: contemporary western European immigration policies and the politics of multiculturalism. *Diaspora* **4**(1): 89–103

Manning C 1995 Approaching the turning point? Labor market change under Indonesia's new order. *The Developing Economies* **33**(1): 52–81

Martin P L 1991 *The unfinished story: Turkish labour migration to Western Europe*. Geneva, International Labour Office

Martin P L 1994 Epilogue: reducing emigration pressure: what role can foreign aid play? In Böhning and Schloeter-Paredes 1994: 241–53

Massey D S 1988 International migration and economic development in comparative perspective. *Population and Development Review* **14**(3): 383–414

Massey D S 1995 The new immigration and ethnicity in the United States. *Population and Development Review* **21**(3): 631–52

Massey D S, Alarcon R, Durand J, Gonzalez H 1987 *Return to Aztlan: the social process of international migration from western Mexico*. Berkeley and Los Angeles, University of California Press

Massey D S, Arango J, Hugo G, Kouaouci A, Pellegrino A, Taylor J E 1993 Theories of international migration: review and appraisal. *Population and Development Review* **19**(3): 431–66

Massey D S, Arango J, Hugo G, Kouaouci A, Pellegrino A, Taylor J E 1994 An evaluation of international migration theory: the North American case. *Population and Development Review* **20**(4): 699–751

Meillassoux C 1981 *Maidens, meal and money*. Cambridge, Cambridge University Press

Miles R 1987 *Capitalism and unfree labour: anomaly or necessity?* London, Tavistock

Miller K A 1985 *Emigrants and exiles: Ireland and the Irish exodus to North America*. New York, Oxford University Press

Mitchneck B, Plane D 1995 Migration patterns during a period of political and economic shocks in the former Soviet Union: a case study of Yaroslavl' Oblast. *Professional Geographer* 47(1): 17–30

Moch L P 1992 *Moving Europeans: migration in Western Europe since 1650*. Bloomington, Indiana University Press

Mosher W D 1980a Demographic responses and demographic transitions: a case study of Sweden. *Demography* 17(4): 395–412

Mosher W D 1980b The theory of change and response: an application to Puerto Rico, 1940 to 1970. *Population Studies* 34(1): 45–58

Myers R H 1995 Taiwan (China). Economy. In *The Far East and Australasia 1995*, 26th edn. London, Europa Publications: 250–4

Nair P R G 1986 India. In Gunatilleke 1986: 66–109

Nair P R G 1989 Incidence, impact and implications of migration to the Middle East from Kerala (India). In R Amjad (ed.) *To the Gulf and back: studies on the economic impact of Asian labour migration*. New Delhi, International Labour Organization, Asian Employment Programme (ARTEP): 344–64

Nelson J M 1979 *Access to power: politics and the urban poor in developing nations*. Princeton, Princeton University Press

Nelson N 1992 The women who have left and those who have stayed behind: rural–urban migration in central and western Kenya. In Chant 1992: 109–38

Noble G R 1982 Epidemiological and clinical aspects of influenza. In A S Beare (ed.) *Basic applied influenza research*. Boca Raton, Fla., CRC Press: 11–45

Nugent W 1992 *Crossings: the great transatlantic migrations, 1870–1914*. Bloomington, Indiana University Press

Ogawa N 1986 *Internal migration in Japanese postwar development*. Tokyo, Nihon University, Population Research Institute Research Paper Series No. 33

Ogawa N, Suits D B 1981 Population change and economic development: lessons from the Japanese experience, 1885–1920. Tokyo, Nihon University, Population Research Institute Research Paper Series No. 2

Ohkawa K 1983 Japan's development: a model for less-developed countries. *Asian Development Review* 1(2): 45–57

Ohmae K 1995 *The end of the nation state: the rise of regional economies*. New York, The Free Press

Omer-Cooper J D 1995 South Africa: recent history. In *Africa south of the Sahara 1995*, 24th edn. London, Europa Publications: 841–50

Omran A R 1971 The epidemiologic transition: a theory of the epidemiology of population change. *Milbank Memorial Fund Quarterly* 49(4): 509–38

O'Sullivan P (ed.) 1992 *Patterns of migration*. Volume 1 of the series entitled *The Irish world wide: history, heritage, identity*. Leicester, Leicester University Press

Oucho J O 1995 Emigration dynamics in eastern African countries. *International Migration* 33(3/4): 391–434

Overholt W H 1993 *China: the next economic superpower*. London, Weidenfeld and Nicolson

Pang E F 1991 Labour migration workers in Singapore: policies, trends and implications. *Regional Development Dialogue* 12(3): 22–34

Pang E F, Lim L 1982 Foreign labor and economic development in Singapore. *International Migration Review* **16**(3): 548–76

Pannell C W, Ma L J C 1983 *China: the geography of development and modernization.* London, Edward Arnold

Papademetriou D G, Martin P L (eds) 1991 *The unsettled relationship: labour migration and economic development.* New York, Greenwood Press

Park C B, Cho N-H 1995 Consequences of son preference in a low-fertility society: imbalance of the sex ratio at birth in Korea. *Population and Development Review* **21**(1): 59–84

Parnwell M 1993 *Population movements and the Third World.* London, Routledge

Passel J S, Woodrow K A 1987 Change in the undocumented alien population in the United States 1979–1983. *International Migration Review* **21**(4): 1304–34

Patterson O 1978 Migration in Caribbean societies: socioeconomic and symbolic resource. In W H McNeill and R S Adams (eds) *Human migration: patterns and policies.* Bloomington, Indiana University Press: 106–45

Peach C 1995 Anglophone Caribbean migration to the USA and Canada. In Cohen 1995a: 245–7

Pérez-López J, Díaz-Briquets S 1990 Labor migration and offshore assembly in the socialist world: the Cuban experience. *Population and Development Review* **16**(2): 273–99

Pessar P R 1991 Caribbean emigration and development. In Papademetriou and Martin 1991: 201–10

Petersen W 1975 *Population*, 3rd edn. New York, Macmillan

Pillai P, Yusof Z A 1996 Malaysia. Paper presented at the meeting on International Migration and the Labour Market in Asia: National Policies and Regional Co-operation. Tokyo, Japan Institute of Labour, 1–2 February

Pope D, Withers G 1993 Do migrants rob jobs? Lessons of Australian history, 1861–1991. *The Journal of Economic History* **53**(4): 719–42

Portes A, Rumbaut R G 1990 *Immigrant America: a portrait.* Berkeley, University of California Press

Potter R G, Kobrin F E 1982 Some effects of spouse separation on fertility. *Demography* **19**(1): 79–95

Potts D 1995 Shall we go home? Increasing urban poverty in African cities and migration processes. *The Geographical Journal* **161**(3): 245–64

Potts L 1990 *The world labour market: a history of migration.* London, Zed Books

Price C A 1974 *The great white walls are built: restrictive immigration to North America and Australasia 1836–1888.* Canberra, Australian National University Press

Prothero R M 1994 Forced movements of population and health hazards in tropical Africa. *International Journal of Epidemiology* **23**(4): 657–64

Prothero R M 1995 Malaria in the nineties. *Geography* **80**(4): 411–14

Pryor R J 1981 Integrating international and internal migration theories.

In M M Kritz, C B Keely and S M Tomasi (eds) *Global trends in migration: theory and research on international population movements.* Staten Island, Center for Migration Studies: 110–29

Pryor R J 1982 Population redistribution, the demographic and mobility transitions. In J I Clarke and L A Kosiński (eds) *Redistribution of population in Africa.* London, Heinemann: 25–30

Ravenstein E G 1885 The laws of migration. *Journal of the Statistical Society* **48**: 167–227

Ravenstein E G 1889 The laws of migration. *Journal of the Statistical Society* **52**: 214–301

Redding S G 1990 *The spirit of Chinese capitalism.* Berlin, Walter de Gruyter

Redford A 1976 *Labour migration in England 1800–1850*, 3rd edn. revised and edited by W H Chaloner. Manchester, Manchester University Press

Reid A 1988 *Southeast Asia in the age of commerce 1450–1680.* Volume 1: *The lands below the winds.* New Haven, Yale University Press

Rhode B 1993 Brain drain, brain gain, brain waste: reflections on the emigration of highly educated and scientific personnel from Eastern Europe. In King 1993a: 228–45

Ricca S 1989 *International migration in Africa: legal and administrative aspects.* Geneva, International Labour Office

Richmond A H 1994 *Global apartheid: refugees, racism, and the new world order.* Toronto, Oxford University Press

Rimmer P 1994 Regional economic integration in Pacific Asia. *Environment and Planning A* **26**(11): 1731–59

Robinson V 1995 The migration of East African Asians to the UK. In Cohen 1995a: 331–6

ROC 1990 *Statistical yearbook of the Republic of China 1990.* Taipei

Rogers A 1982 Sources of urban population growth and urbanization, 1950–2000: a demographic accounting. *Economic Development and Cultural Change* **30**(3): 483–506

Rogers A (ed.) 1992 *Elderly migration and population redistribution: a comparative study.* London, Belhaven

Rowland R N 1983 The growth of large cities in the USSR: policies and trends 1959–1979. *Urban Geography* **4**(3): 258–79

Rowland R 1993 Regional migration in the former Soviet Union during the 1980s: the resurgence of European regions. In King 1993a: 152–74

Rozman G 1990 East Asian urbanization in the nineteenth century: comparisons with Europe. In A D van der Woude, J de Vries and A Hayami (eds) *Urbanization in history.* Oxford, Clarendon Press

Russell S S 1986 Remittances from international migration: a review in perspective. *World Development* **14**(6): 677–96

Russell S S 1992 Migrant remittances and development. *International Migration* **30**(3/4): 267–87

Russell S S, Jacobsen K, Stanley W D 1990 International migration and development in sub-Saharan Africa. World Bank Discussion Papers, 2 vols., Nos. 101 and 102. Washington DC, World Bank

Russell S S, Teitelbaum M S 1992 International migration and international trade. World Bank Discussion Papers, No. 160. Washington DC, World Bank

Said E W 1993 *Culture and imperialism*. London, Chatto and Windus

Salt J 1993 External international migration. In D Noin and R Woods (eds) *The changing population of Europe*. Oxford, Blackwell: 185–97

Salt J, Clout H (eds) 1976 *Migration in post-war Europe: geographical essays*. London, Oxford University Press

Sanguin A-L 1994 Les réseaux des diasporas. *Cahiers de Géographie du Québec* **38**(105): 495–8

Sassen S 1988 *The mobility of labour and capital*. Cambridge, Cambridge University Press

Sassen S 1991 *The global city: New York, London, Tokyo*. Princeton, Princeton University Press

Sassen S 1995 Labour mobility and migration policy: lessons from Japan and the U.S. In B Unger and F Van Waarden (eds) *Convergence or diversity: internationalization and economic policy response*. Aldershot, Avebury: 108–31

Saw S-H 1991 Population growth and control. In E C T Chew and E Lee (eds) *A history of Singapore*. Singapore, Oxford University Press

Schaeffer P V 1993 A definition of migration pressure based on demand theory. *International Migration* **31**(1): 43–72

Scott J C 1987 *Weapons of the weak: everyday forms of peasant resistance*. New Haven, Yale University Press

Segal R 1994 *The black diaspora*. London, Faber and Faber

Selya R M 1995 Taiwan as a service economy. *Geoforum* **25**(3): 305–22

Shafir G 1995 Zionist immigration and colonization in Palestine until 1948. In Cohen 1995a: 405–9

Shah N M 1994 Arab labour migration: a review of trends and issues. *International Migration* **32**(1): 3–28

Shah N M 1995 Structural changes in the receiving country and future labour migration: the case of Kuwait. *International Migration Review* **29**(4): 1000–22

Shimpo M 1995 Indentured migrants from Japan. In Cohen 1995a: 48–50

Shortridge K F 1995 The next pandemic influenza virus? *The Lancet* **346**(8984): 1210–12

Simmons A, Díaz-Briquets S, Laquian A A 1977 *Social change and internal migration: a review of research findings from Africa, Asia, and Latin America*. Ottawa, International Development Research Centre

Simmons A B, Guengant J P 1992 Caribbean exodus and the world system. In Kritz, Lim and Zlotnik 1992: 94–114

Simon D 1992 *Cities, capital and development: African cities in the world economy*. London, Belhaven

Simon J L 1981 *The ultimate resource*. Oxford, Martin Robertson

Sinn E 1995 Emigration from Hong Kong before 1941: general trends. In R Skeldon (ed.) *Emigration from Hong Kong: tendencies and impacts*. Hong Kong, The Chinese University Press: 11–34

Skeldon R 1976 Regional associations and population migration: an interpretation. *Urban Anthropology* **5**(3): 233–52

Skeldon R 1985 Population pressure, migration and socio-economic change in mountainous environments: regions of refuge in comparative perspective. *Mountain Research and Development* 5(3): 233–50

Skeldon R 1986 On migration patterns in India during the 1970s. *Population and Development Review* 12(4): 759–79

Skeldon R 1987a Migration and the population census in Asia and the Pacific: issues, questions and debate. *International Migration Review* 21(4): 1074–1100

Skeldon R 1987b Protest, peasants and the proletariat: the circular migrant as catalyst. In R Ghose (ed.) *Protest movements in South and South-east Asia.* Hong Kong Centre of Asian Studies: 3–18

Skeldon R 1990 *Population mobility in developing countries: a reinterpretation.* London, Belhaven

Skeldon R 1990/1 Emigration and the future of Hong Kong. *Pacific Affairs* 63(4): 500–23

Skeldon R 1992 On mobility and fertility transitions in East and Southeast Asia. *Asian and Pacific Migration Journal* 1(2): 220–49

Skeldon R (ed.) 1994 *Reluctant exiles? Migration from Hong Kong and the new overseas Chinese.* New York, M E Sharpe and Hong Kong, Hong Kong University Press

Skeldon R 1995a Immigration and population issues. In Y-L Cheung and S M H Sze *The other Hong Kong report 1995.* Hong Kong, The Chinese University Press

Skeldon R 1995b The challenge facing migration research: a case for greater awareness. *Progress in Human Geography* 19(1): 91–6

Skeldon R 1996 Migration from China. *Journal of International Affairs* 49(2): 434–55

Skeldon R 1997a Hong Kong and south China growth linkages. In D Campbell, A Parisotto and A Verma (eds) *Regionalization and labour market interdependence in East and Southeast Asia.* London, Macmillan

Skeldon R 1997b Hong Kong communities overseas. In J M Brown and R Foot (eds) *Hong Kong transitions, 1842–1997.* Basingstoke, Macmillan

Skeldon R 1997c Migrants on a global stage: the Chinese. In P J Rimmer (ed.) *Pacific Rim development: integration and globalisation in the Asia-Pacific economy.* Sydney, Allen and Unwin: 222–39

Sklair L 1993 *Assembling for development: the maquila industry in Mexico and the United States,* 2nd edn. San Diego, Center for US–Mexican Studies, University of California at San Diego

Smart J 1994 Business immigration to Canada: deception and exploitation. In Skeldon 1994: 98–119

Smil V 1992 Environmental change as a source of conflict and economic losses in China. Occasional Paper Series: environmental change and acute conflict. Cambridge, Mass., American Academy of Arts and Sciences

Smil V 1993 *China's environmental crisis: an inquiry into the limits of national development.* New York, M E Sharpe

SOPEMI 1994 *Annual report 1993: trends in international migration.* Paris, OECD

South Centre 1993 *Facing the challenge: responses to the report of the South Commission.* London, Zed Books

Sowell T 1996 *Migrations and cultures: a world view.* New York, Basic Books

Speare A 1974 Urbanization and migration in Taiwan. *Economic Development and Cultural Change* 22(2): 302–17

Stahl C, Habib A 1991 Emigration and development in south and southeast Asia. In Papademetriou and Martin 1991: 163–79

Stalker P 1994 *The work of strangers: a survey of international labour migration.* Geneva, International Labour Office

Stark O 1991 *The migration of labor.* Oxford, Blackwell

Storper M, Walker R 1989 *The capitalist imperative: territory, technology, and industrial growth.* Oxford, Blackwell

Straubhaar T 1993 Migration pressure. *International Migration* 31(1): 5–41

Streeton P 1993 The special problems of small countries. *World Development* 21(2): 197–202

Suhrke A 1993 Pressure points: environmental degradation, migration and conflict. Occasional Paper Series: environmental change and acute conflict. Cambridge, Mass., American Academy of Arts and Sciences

Tan E S, Chiew S K 1995 Emigration orientation and propensity: the Singapore case. In J H Ong, K B Chan and S B Chew (eds) *Crossing borders: transmigration in Asia Pacific.* Singapore, Prentice Hall: 239–58

Taylor J, Bell M 1996 Population mobility and indigenous peoples: the view from Australia. *International Journal of Population Geography* 2(2): 153–69

Teitelbaum M S, Winter J M 1985 *The fear of population decline.* Orlando, Fla., Academic Press

Thant M, Tang M, Kakazu H 1994 *Growth triangles in Asia: a new approach to regional economic cooperation.* Hong Kong, Oxford University Press

Theroux P 1995 *The pillars of Hercules: a grand tour of the Mediterranean.* London, Hamish Hamilton

Thiam B 1994 Environmental impact on migration and on the spatial redistribution of the population. In *Population, environment and development.* New York, United Nations: 175–85

Thio E 1991 The Syonan years, 1942–1945. In E C T Chew and E Lee (eds) *A history of Singapore.* Singapore, Oxford University Press

Thomas B 1954 *Migration and economic growth: a study of Great Britain and the Atlantic economy.* Cambridge, Cambridge University Press

Thomas-Hope E M 1992 *Explanation in Caribbean migration.* Basingstoke, Macmillan

Tilly C 1984 *Big structures, large processes, huge comparisons.* New York, Russell Sage Foundation

Tinker H 1974 *A new system of slavery: the export of Indian labour overseas 1830–1920.* London, Oxford University Press

Tinker H 1977 *The banyan tree: overseas emigrants from India, Pakistan and Bangladesh.* Oxford, Oxford University Press

Tirtosudarmo R 1990 Transmigration policy and national development plans in Indonesia (1969–88). Working paper 90/10. Canberra, National Centre for Development Studies, Australian National University

Todaro M P 1969 A model of labor migration and urban unemployment in less developed countries. *The American Economic Review* **59**(1): 138–48

Todaro M P 1976 *Internal migration in developing countries*. Geneva, International Labour Office

Todaro M P 1994 *Economic development*, 5th edn. New York, Longman

Tomasi L, Miller M J (eds) 1992 The new Europe and international migration. *International Migration Review* **26**(2): special issue

Topley M 1964 Capital, saving and credit among indigenous rice farmers and immigrant vegetable farmers in Hong Kong's New Territories. In R Firth and B S Yamey (eds) *Capital saving and credit in peasant societies: studies from Asia, Oceania, the Caribbean and Middle America*. Chicago, Aldine: 157–86

Tribalat M 1992 Chronique de l'immigration. *Population* **47**(1): 153–90

Tribalat M 1996 Chronique de l'immigration. *Population* **51**(1): 141–91

Tsai H-C 1988 A study on the migration of students from Taiwan to the United States: a summary report. *Journal of Population Studies* **12**: 91–120

Tsay C L 1995 Data on international migration to Taiwan. *Asian and Pacific Migration Journal* **4**(4): 613–19

Tsokhas K 1994 Immigration and unemployment in Australia. *International Migration* **32**(3): 445–66

Tsuya N O, Kuroda T 1992 Japan: the slowing of urbanization and metropolitan concentration. In A G Champion (ed.) *Counterurbanization: the changing pace and nature of population deconcentration*. London, Edward Arnold: 207–29

Turnbull C M 1982 *A history of Singapore 1819–1975*, 2nd edn. Singapore, Oxford University Press

Turner F J 1894 The significance of the frontier in American history. *Annual report of the American Historical Association for the year 1893*. Washington DC, Government Printing Office

Turner H A 1980 *The last colony: but whose? A study of the labour market and labour relations in Hong Kong*. Cambridge, Cambridge University Press

Turner V, Turner E 1978 *Image and pilgrimage in Christian culture: anthropological perspectives*. New York, Columbia University Press

United Nations 1991 *The challenge of free economic zones in central and eastern Europe: international perspectives*. New York, United Nations Centre on Transnational Corporations

United Nations 1994 Population distribution and migration. Proceedings of the United Nations Expert Meeting on Population Distribution and Migration, Santa Cruz, Bolivia, 18–22 January 1993. New York

United Nations 1995a *Human development report 1995*. New York, Oxford University Press

United Nations 1995b *International migration policies 1995* (wall chart). New York, Department of Economic and Social Information and Policy Analysis

United Nations 1995c *World investment report 1995: transnational corporations and competitiveness.* New York and Geneva, United Nations Conference on Trade and Development

United Nations 1996 *World population monitoring 1993 with a special report on refugees.* New York, Department for Economic and Social Information and Policy Analysis

van den Broek J (ed.) 1996 *The economics of labour migration.* Cheltenham, Edward Elgar

Vasquez N D 1992 Economic and social impact of labor migration. In Battistella and Paganoni 1992: 41–67

Vasquez N D, Tumbaga L C, Cruz-Soriano M 1995 *Tracer study on Filipino domestic helpers abroad.* Geneva, International Organization for Migration

Ventura R 1992 *Underground in Japan.* London, Jonathan Cape

Waldinger R 1992 Taking care of the guests: the impact of immigration on services – an industry case study. *International Journal of Urban and Regional Research* **16**(1): 96–113.

Waldinger R 1993 The two sides of ethnic entrepreneurship: reply to Bonacich. *International Migration Review* **27**(3): 692–701

Wallerstein I 1974 *The modern world system.* Vol. 1: *Capitalist agriculture and the origins of the European world economy in the sixteenth century.* New York, Academic Press

Wallerstein I 1980 *The modern world system.* Vol. 2: *Mercantilism and the consolidation of the European world economy 1600–1750.* New York, Academic Press

Wallerstein I 1989 *The modern world system.* Vol. 3: *Second era of great expansion of the capitalist world economy 1730–1840s.* New York, Academic Press

Wallerstein I 1993 Wise, but not tough, or is it correct, but not wise? In The South Centre, *Facing the challenge: responses to the report of the South Commission.* London, Zed Books: 117–21

Walsh A C 1992 The status of circular migration in the evolution of Melanesian towns: an attempt at explanation. *Asian and Pacific Population Journal* **1**(2): 196–219

Wang G (ed.) 1997 *Global history and migration.* Boulder, Westview Press

Ward R G 1989 Earth's empty quarter? The Pacific islands in a Pacific century. *The Geographical Journal* **155**(2): 235–46

Watanabe S 1994 The Lewisian turning point and international migration: the case of Japan. *Asian and Pacific Migration Journal* **3**(1): 119–47

Watson, J L 1975 *Emigration and the Chinese lineage: the Mans in Hong Kong and London.* Berkeley, University of California Press

Weber E 1977 *Peasants into Frenchmen: the modernization of rural France 1870–1914.* London, Chatto and Windus

Weiner Michael 1994 *Race and migration in imperial Japan.* London, Routledge

Weiner Myron (ed.) 1993 *International migration and security.* Boulder, Westview Press

Weiner Myron 1995 *The global migration crisis: challenge to states and to human rights.* New York, Harper Collins

Weiner Myron 1996 Nations without borders: the gifts of folk gone abroad. *Foreign Affairs* 75(2): 128–34

Weiss L, Hobson J M 1995 *States and economic development: a comparative historical analysis.* Cambridge, Polity Press

Weniger B G, Limpakarnjanarat K, Ungchusak K, Thanprasertsuk S, Choopanya K, Vanichseni S, Uneklabh T, Thongcharoen P, Wasi C 1991 The epidemiology of HIV infection and AIDS in Thailand. *AIDS* 5(supplement 2): S71–S85

White L T 1994 Migration and politics on the Shanghai delta. *Issues and Studies* 30(9): 63–94

White P 1995 Geography, literature and migration. In King, Connell and White 1995: 1–19

Williamson J G 1988 Migration and urbanization. In H Chenery and T N Srinivasan (eds) *Handbook of development economics,* Vol. 1. Amsterdam, Elsevier Science Publishers: 425–65

Wolpert J 1965 Behavioral aspects of the decision to migrate. *Papers, Regional Science Association* 15: 159–72

Wong D 1996 Men who built Singapore: Thai workers in the construction industry. Draft report submitted to the Workshop on Labor Migrants and Economic Development in Mainland Southeast Asia, Bangkok, 23–25 May. Asian Research Center for Migration, Institute of Asian Studies, Chulalongkorn University

Wong S-L 1992 Emigration and stability in Hong Kong. *Asian Survey* 32(10): 918–33

Wood H 1984 *Third class ticket.* Harmondsworth, Penguin Books

Wooden M 1994 The economic impact of immigration. In M Wooden, R Holton, G Hugo and J Sloan (eds) *Australian immigration: a survey of the issues,* 2nd edn. Canberra, Australian Government Publishing Service: 111–57

Woodrow K A 1990 Emigration from the United States using multiplicity surveys. Paper presented at the Annual Meeting of the Population Association of America, Toronto, 3–5 May

World Bank 1990 *Poverty: world development report 1990.* New York, Oxford University Press

World Bank 1993 *The East Asian miracle: economic growth and public policy.* New York, Oxford University Press

World Bank 1995 *World development report 1995.* New York, Oxford University Press

Wright J J 1991 *The balancing act: a history of modern Thailand.* Bangkok, Asia Books

WTO 1994 *Yearbook of tourism statistics,* 46th edn. Madrid, World Tourism Organization

Wu C T 1997 Globalization of the Chinese countryside: international capital and the transformation of the Pearl River Delta. In P J Rimmer (ed.), *Pacific Rim development: integration and globalisation in the Asia–Pacific economy.* Sydney, Allen and Unwin: 57–82

Wu Y-L, Wu C H 1980 *Economic development in southeast Asia: the Chinese dimension.* Stanford, Calif., Hoover Institution Press

Yanitsky O, Zaionchkovskaya Z 1984 Soviet sociology relating to rural migrants in cities. *International Social Science Journal* 36(3): 469–85

Yap M T 1991 *Singaporeans overseas: a study of emigrants in Australia and Canada*. Report No. 3. Singapore, The Institute of Policy Studies

Yongyuth C 1996 Thailand. Paper presented at the meeting on International Migration and the Labour Market in Asia: National Policies and Regional Cooperation. Tokyo, Japan Institute of Labour, 1–2 February

Yukawa J 1996 *Migration from the Philippines, 1975–1995: an annotated bibliography*. Quezon City, the Philippines, Scalabrini Migration Center

Zachariah K C 1977 Measurement of internal migration from census data. In A A Brown and E Neuberger (eds) *Internal migration: a comparative perspective*. New York, Academic Press: 121–34

Zachariah K C, Conde J 1981 *Migration in West Africa: demographic aspects*. New York, Oxford University Press

Zelinsky Z 1971 The hypothesis of the mobility transition. *Geographical Review* **61**(2): 219–49

Zeng Y, Tu P, Gu B, Xu Y, Li B, Li Y 1993 Causes and implications of the recent increase in the reported sex ratio at birth in China. *Population and Development Review* **19**(2): 283–302

Zimmermann K F 1995 Tackling the European migration problem. *Journal of Economic Perspectives* **9**(2): 45–62

Zlotnik H 1991 Trends in South to North migration: the perspective from the North. *International Migration* **29**(4): 317–31

Zlotnik H 1992 Empirical identification of international migration systems. In M M Kritz, L L Lim and H Zlotnik (eds) *International migration systems: a global approach*. Oxford, Clarendon Press

Zolberg A R 1978 International migration policies in a changing world system. In W H McNeill and R S Adams (eds) *Human migration: patterns and policies*. Bloomington, Indiana University Press: 241–86

Zolberg A R 1981 Origins of the modern world system: a missing link. *World Politics* **33**: 253–8

Zolberg A R, Suhrke A, Aguayo S 1989 *Escape from violence: conflict and the refugee crisis in the development world*. New York, Oxford University Press

General Index

Geographical Index

Arranged by large geographical area
Countries listed in annexe tables 1 and 2 are not included in the index.

Null Hypothesis - 1%